When Science and Politics Collide

When Science and Politics Collide

The Public Interest at Risk

Robert O. Schneider

 PRAEGER™

An Imprint of ABC-CLIO, LLC

Santa Barbara, California • Denver, Colorado

Library of Congress Cataloging-in-Publication Data

Names: Schneider, Robert O., author.
Title: When science and politics collide : the public interest at risk /
    Robert O. Schneider.
Description: Santa Barbara, California : Praeger, An Imprint of ABC-CLIO, LLC, 2018. |
    Includes bibliographical references and index.
Identifiers: LCCN 2018000408 (print) | LCCN 2018005513 (ebook) | ISBN
    9781440859380 (ebook) | ISBN 9781440859373 (set : alk. paper)
Subjects: LCSH: Science—Political aspects—United States. | Science and
    state—United States.
Classification: LCC Q175.52.U5 (ebook) | LCC Q175.52.U5 S285 2018 (print) |
    DDC 320.6—dc23
LC record available at https://lccn.loc.gov/2018000408

ISBN: 978-1-4408-5937-3 (print)
        978-1-4408-5938-0 (ebook)

22 21 20 19 18   1 2 3 4 5

This book is also available as an eBook.

Praeger
An Imprint of ABC-CLIO, LLC

ABC-CLIO, LLC
130 Cremona Drive, P.O. Box 1911
Santa Barbara, California 93116-1911
www.abc-clio.com

This book is printed on acid-free paper ∞
Manufactured in the United States of America

Contents

CHAPTER ONE

Science and Politics

Introduction

When do science and politics collide? The short answer is, almost always. The reasons for this are many. Science and politics are very different things, and scientists and politicians are animated by very different motivations. We all live in a world where it has become more important than ever to make intelligent decisions driven by the properties of the physical universe. But the relationship between science and politics is, as we shall see, a troubled one. The ways that science and politics interact with each other in the policy-making dialogue often do not serve the public interest, and in some cases they actually result in great harm. This is a problem that requires considerable effort to understand, and until we really understand it, we will not solve it.

Scientific advice to policy makers has featured prominently in American history. But the mixing of politics and science can be a bit like mixing acid and sulfides. The results are often and regrettably toxic. Policy makers have, as we have all no doubt observed during the course of our own lifetimes, become ever more dependent on expert scientific advice as they navigate the most complex issues in a technological age.[1] But it is also true that scientific knowledge and expertise have never been able to command unquestioned authority and public trust. In part, this is due to the selective application of science and its expert advice by policy makers. The political actors often view science as something they can use when they find it "supportive" of the specific goals or purposes they wish to advance. At the same time, where scientific knowledge or expertise is at odds with the preferences of policy makers, it is often very quickly deep-sixed or delegitimized in the (increasingly, I believe) fact-free zone of partisan political discourse.[2] It is also true that science and its conclusions often run contrary to the majority of public opinion and, as such, seem undemocratic or unaccountable to the public and its expressed preferences or beliefs. In either case, the formula for politically explosive conflict is all too easily mixed. To all this may be added the reluctance of scientists to see what they do as having any connection to the political world or, perhaps more precisely, their insistence that they are totally separate from that world and wish to remain so.

The experience of J. Robert Oppenheimer is perhaps a prime example of the inherently uneasy relationship between science and politics. Despite what some regarded as questionable political associations in his past, his expertise as a physicist and his ability to contribute to a project of great national importance during World War II led to his being named director of the Manhattan Project at Los Alamos. Beginning in 1943, Oppenheimer directed nearly 6,000 scientific and military employees in the effort to develop the atomic bomb. His genius and leadership and his knowledge about the potential of atomic energy were beyond question. But when that genius and knowledge led him to believe that the United States needed to be open minded and to be a force for promoting international responsibility with respect to atomic power, he was less well-regarded by the policy elite. After the war, Oppenheimer was appointed chairperson of the newly created United States Atomic Energy Commission. From that distinguished position, he began to lobby for international control of nuclear power and to advocate efforts to prevent nuclear proliferation. He opposed the building of the hydrogen bomb. In fact, he opposed a nuclear-arms race with the Soviet Union. This provoked the ire of politicians and led to the revocation of his security clearance and an end to his effective influence with policy makers.[3]

It could be argued that it was Oppenheimer's political views that were the problem. Both the scientist and the politician would undoubtedly agree with this argument. In other words, it was not his scientific views or his expertise that earned him disfavor. It was the policy he advocated or the political views that deviated from those of the policy elites. One can imagine that both the scientist and the politician might agree that appearing to advocate a political point of view was not his job as a scientist. Yet even purely scientific views may be resisted for political reasons. What are we to say, for example, when scientific expertise reaches peer-reviewed scientific conclusions that come into conflict with the political or policy preferences of the policy makers? Is the scientific view to be expressed only when it leads to a conclusion that is politically satisfactory to the policy maker? Is the political or policy recommendation to be dismissed as mere ideology and the science supporting it to be considered partisan? And is the scientist to be forever banned or given to self-censorship when legitimate scientific knowledge may provide a foundation for raising policy and value questions? Consider that bomb manufacturing served the government's goals and purposes in the postwar world as a new conflict, the Cold War, was born. Might policy makers who disregarded the experts (physicists and other world-class scientists) who questioned those purposes or suggested different purposes have been reckless in their avoidance of important information or analysis? These are vexing questions.

The most vexing of questions may be those that move the conversation in one of two directions. First is the inevitable problem of scientists appearing to engage in policy advocacy, a role not deemed appropriate for them in many professional and political circles. This may lead to the inevitable rejection of the science because the policy being advocated is not popular with policy makers or the public. Second, there is the equally vexing problem of squaring science with a politics that often insists on the validity of scientific viewpoints that are inconsistent with rigorously reviewed and well-accepted scientific conclusions. There are many more such vexing questions in the relationship between science and politics. Perhaps the place to begin sorting things out is by asking and attempting to answer two very basic questions: What is science? What is politics? As we all undoubtedly know, they are two very different things. Though they sometimes participate in the same discussion of important issues of public concern, they do not speak the same language or serve the same purposes. Frequently, and very troublingly, where the broader public interest is concerned, they do not even occupy the same reality.

What Is Science?

Before delving into what science is, let us begin with an observation. It is factually true that scientists tend to be politically liberal in the common American usage of that ideological term.[4] Very few scientists identify as conservative or Republican (see Tables 1.1 and 1.2). In part, this is because scientists accept evidence that supports things proven to be scientifically true that contemporary Republicans and conservatives often do not believe in or accept, including things like evolution and anthropogenic climate change, to name just two. Today, scientists are "scientifically" and understandably at odds with many leading Republican members of the U.S. Senate and House of Representatives who are deniers of climate change and evolution.

Table 1.1   Partisan Affiliations of Public and Scientists

Partisan Affiliation	Public	Scientists
Democrat	35%	55%
Republican	23%	6%
Independent	34%	32%
Other	4%	4%

Source: Pew Research Center, www.people-press.org/files/legacy/528-52.gif

Congress member Lamar Smith, Republican from Texas and chair of the House Committee on Science, Space, and Technology, recently began to champion increased political oversight of the National Science Foundation's grant-making process. While not an alarming thing in itself, as Congress does have oversight authority and responsibility, some felt the intent was primarily political and partisan in nature.[5] When Representative Joseph Barton, Republican from Texas and chair of the House Energy and Commerce Committee, subpoenaed climate researcher Michael E. Mann and his colleagues, asking for the raw data and computer codes from all their research, it struck many that scientific peer review would now be replaced by partisan political review of science.[6] Both Barton and Smith are leading climate-change deniers, and their efforts have been characterized by some as an attempt to intimidate scientists. This raises the disturbing prospect of politicians opening investigations against any scientists who make them feel uncomfortable because their science does not conform to the partisan ideological beliefs of the policy maker. I'm sure these Republican lawmakers fear that science has a liberal bias.

While such tension between elected policy makers and scientists appears to be political on both sides—after all, the scientists are more liberal than conservative—the assumption of a liberal bias by science is a fictive concept. Many American citizens are inclined, in no small measure due to the partisan criticism of science itself as partisan or biased, to spurn scientific consensus and to be unreasonably suspicious of research. But, here again, it can be argued that there is nothing in science, *as science,* that contributes to this reaction. Rather, it is most often due to the inability of science to be heard above the echoes of doubt manufactured by partisans and policy makers who do not accept scientific conclusions that disagree with their beliefs or preferences. Let us examine this in some detail.

We have seen in recent years in the United States that the federal government has expanded its support for scientific research and development,

Table 1.2   Ideological Comparison of Public and Scientists

Ideology	Public	Scientists
Conservative	37%	9%
Moderate	38%	35%
Liberal	20%	52%
Very Liberal	5%	4%

Source: Pew Research Center, www.people-press.org/files/legacy/528-52.gif

especially in the life sciences. But this has been accompanied by partisan charges that science itself has become politicized and that scientists themselves are politically biased in their work. This has resulted in a corresponding political and public disregard for scientific evidence that has greatly heightened tension between the scientific and political communities.

As we have witnessed the many highly publicized debates about things like stem-cell research, vaccines, evolution, and anthropogenic climate change, we have heard many competing and irreconcilable claims made by scientists, special interests, and members of Congress. Strategically manipulated questions about the reliability or soundness of the science as it pertains to the policy debate at hand, together with pseudoscientific policy reports crafted by ideological think tanks and presented to skew the discussion, have no doubt caused many to think that science itself is not reliable. We see a U.S. senator holding a snowball on the floor of the Senate and calling climate change a hoax. Incredibly, some regard this as a legitimate response to the testimony of a climate scientist presenting the latest peer-reviewed research. Of course, special interests paying for pseudoscience that supports their interests intentionally muddy the scientific waters to their advantage. Also, it must be noted in this context that legitimate scientific disagreement and unresolved or unanswered questions are manipulated to provide all the more cover for those who wish to reject what science does, in fact, have to say to us. Indeed, one member of Congress told a scientist testifying at a hearing, "You have your science, I have mine."[7] What we witness on a daily basis in the halls of Congress and in state legislative bodies across the nation is the persistent demotion of rigorous and peer-reviewed science to the level of political opinion or mere ideology. This confuses the public about science and what it actually is.

It is no small thing to understand what science actually is. Knowing what science does, knowing how the scientific method works, cannot be taken for granted, however. For all the world-leading scientific inquiry conducted by American scientists, Americans aren't too well versed in science or much interested in it. American primary and secondary school students lag behind their counterparts around the world in terms of test scores. Relatively few go into the sciences. Both China and Japan produce a larger percentage of science graduates than the United States. One consequence of this is that important and growing American industries in science and technology have many more job openings than we have workers who can do the job.[8] It is all too common for even educated people to be ignorant of science. We have all heard well-educated people say, "Oh, I'm not very good at math" or "I never took physics." Yet, as we have said at the outset, we all live in a world where we must make intelligent decisions driven by the properties of the physical universe.

It can perhaps be agreed, even by those reluctant to identify a policy or advocacy role for science, that if scientists do not offer their expert knowledge to the policy process, or if policy makers do not incorporate this expertise into their analysis, we risk making ill-informed policy decisions on technically complex issues. But many would also no doubt contend that if scientists play too large a role in defining the issues and their solutions, we risk the weakening of democratic governance and the rise of technocratic control. Incorporating science into the democratic policy process would be easy if there were a firm boundary between the factual, objective world of science and the value-laden arena of political debate and decision making. Ideally, given such a boundary, scientific facts could serve as neutral information to guide public policy. But such an absolute boundary does not exist in practice. Science cannot help but incorporate values into its assessments of problems and its identification of solutions. Science is, in fact, political to the extent that *its findings are relevant to the issues we must decide on*. Scientific findings may cause us to question long-held beliefs and policy preferences. Even the negotiation over what science is necessary to inform public policy takes place in tandem with negotiation over competing values that are advanced or hindered by the alternative policy solutions themselves. But understanding what science actually is might improve our chances of easing some of the tension inherent between science and politics in these negotiations.

Science, we must note, is very deliberate. It is equally important to note that in science, as in life itself, 100 percent certainty is unattainable. But science is not about the assertion of certainty; science is about evidence. Science gathers evidence and subjects it to legitimate and methodologically rigorous analysis. This analysis provides the basis for reaching conclusions, and the reliability of these conclusions must be verified by further observation and testing. All scientific conclusions, to the extent they may be said to be reliable, must meet the test of rigorous peer review by other scientists. Science is disciplined inquiry, and its work is aimed at a refined understanding and an enhanced explanation of the properties of the physical universe. Its methods and its rules of evidence are strict. It does not yet have all the answers to all the questions that matter, but it has advanced our knowledge, rendered the world knowable, and increased our capacity to predict much about it. At its best, science reaches an expert consensus that provides us with the best foundation for what we can safely say is scientific truth. Of course, scientific truth is not an end point. It is an emerging thing that is always being refined and extended. It is an active phenomenon. Science is a process. If it stops, so do we.

When one thinks about science, one might think of knowledge. One might think of an ever-better understanding of our natural world, and of our constructed world, and what knowing more might allow us to do in the future. One might think, for example, of our need to better anticipate that future and to plan for a world of super global competition. This is a world where smart companies lead, new industries and sources of wealth emerge, and a differently skilled workforce is necessary. Science in this context can be said to be a necessity for success. One might also think of the evidence that is relentlessly accumulating that we humans and our actions are changing our planet. Science provides evidence of the many challenges this may pose to our health and survivability. It convincingly articulates the need for us to better understand and improve our relationship with the natural systems we rely on. And it presents us with options to manage, mitigate, or adapt to these challenges. One might think of science as a necessity for human health, well-being, and progress. It is the human quest to thrive and survive. Without science, neither is a possibility.

Science is an ongoing analytical activity, not the assertion of undeniable truths to be forever embraced and codified. It is an ongoing quest for more confirming or disconfirming data. It constantly questions itself with the perfect awareness that its work is never completed and its journey never ending. Scientific research produces evidence that is robust, and, to the extent it is effectively verified, it is the closest approximation to truth attainable about the world and the universe. If, over time, repeated study and testing by the community of scientists produces the evidence, the result is something called scientific knowledge. This means that sufficient proof has met all the tests of reliability, and, on the basis of that painstaking effort, there is a consensus of expert opinion. Now a single expert, or a small subset of experts, may dissent from the consensus or be skeptical regarding it, but for that dissent or skepticism to be legitimate, it must do more than deny or reject the consensus; it must strive to produce the research *and the evidence* that either disproves the consensus or provides the basis for a new or better consensus. For skeptics also, science is inquiry, and it is all about evidence.

Now it is perhaps still appropriate to ask, even after this idealistic discussion of science, how do we know we can trust the scientists? With respect to their scientific work, that is, science done properly and meeting the test of peer review, we can trust scientists. Why? In no small part because science properly conducted is inevitably going to discover and reveal erroneous findings or sloppy conclusions. The peer review of professional publications is generally very reliable. It may be that scientists are no more or less trustworthy than any other group of people, but science is, after all is said and done, about reliable conclusions and testable predictions. Scientists cannot hide behind untrustworthy results forever because sooner or later someone will reveal their work as false. So, if anything, there may be more than a little support for the view that scientists may *have to be* slightly more trustworthy than other people. The way science works, they are professionally incentivized to be such. That said, some scientists do more than science in the purely academic sense.

Throughout our nation's history, corporations and political or ideological think tanks have supported scientific research. Most of the time, it is safe to assume that they are funding research that will support their economic or political interests. It may also be the case that the scientists being funded know they are expected to produce the results the funder wants if they wish to continue to receive funding. Some would say that the same dynamic may be at work with government-funded research, though that is not as likely as some partisans wish to make it seem. Still, it may well be true that incentives other than sound science may motivate some researchers. One key indicator regarding the soundness or trustworthiness of scientific research is where and how the research is presented. Research appearing in legitimate scientific journals and publications has usually met the test of reliable peer review and may at least be assumed to be sound or valid. Research published outside of that arena may not have been peer reviewed at all and may, in fact, have been published to sway public opinion and not scientific opinion.

An example regarding the publication of materials that question or deny the scientific consensus about anthropogenic (i.e., human-caused) climate change is an interesting case in point. One study found that from 1972 to 2005, 92 percent of climate-change denial or contrarian books originated from conservative think tanks. These think tanks have political agendas, not scientific ones. They are often ground zero for misinformation about climate change and the science that explains it. Their purpose is to influence public opinion and to be of use to conservative media and politicians who oppose climate policy.[9] This effort to distort science in order to impact public opinion works all too well.

Climate-change denial persists in the face of accumulating evidence and strengthening scientific consensus. It is true that as the science has become more certain, public acceptance of the conclusions of climate science has increased slightly. Indeed, as the science has become more robust, more Americans are willing to trust the scientists. But there is a segment of the American population for which the exact opposite is true. Among the most conservative free-market advocates and others who form the conservative political base in the United States, as the scientific consensus has become more robust, trust in scientists fell lower. This response was driven, according to various empirical studies, by the expectation that climate scientists were falsifying evidence to support human-caused climate change.[10] This is a thought pattern that, psychologists tell us, is associated with conspiratorial thinking. Conspiratorial thinking is immune to new evidence. In fact, any new evidence that disproves the conspiracy is immediately viewed as part of the conspiracy.[11] Now, with this observation, we are well beyond science and ready for a brief discussion of politics.

What Is Politics?

It can safely be said that Americans are defined by their radically individualistic orientations. They are not without social awareness or conscience, but their lives see them trapped between two incompatible conceptions of political life. The first is an ideal conception. It is perhaps one inspired by our textbook notions of representative government and our early socialization into our culture. The ideal conception says that, in our republic, consensus occurs spontaneously between free like-minded individuals. Next, we have the realistic conception that sees politics as a battle between conflicting interests or parties to advance or achieve their selfish interests in public policy.[12] In this battle, we do not exactly see citizenship as a habit of heart that allows us to see our differing interests as interrelated. We do not, as a rule, quite see how achieving a good for someone else may be connected to our own good. This may seem harsh, and we can all identify idealistic discussions and motivations in our communities, perhaps even in our national political arena, but these idealistic expressions are atypical or counter to most of our actual experiences as Americans.

American individualism characterizes the way Americans live their lives. Most average middle-class Americans are wrapped up in microsocial relations with family and like-minded friends. They avoid and/or ignore macrosocial relations with "Big Government" and "Big Business."[13] The irony, of course, is that the microsocial relations are influenced profoundly by the macrosocial institutions. People may feel in control of their micro-social lives and yet be totally unaware of how their lives are being manipulated and/or impacted by the events of the macrosocial world. They tend not to participate in the larger world except for occasionally (the frequency is in decline, in fact) watching the news media and sometimes voting in elections.[14] This is the American cultural reality. Alexis de Tocqueville observed it almost 200 years ago when he noted that Americans were disposed to withdraw from society as a whole and retreat into a circle of family and friends.

Despite our individualism, and our preference for the microsocial, American society has many identifiable groups that promote political and policy interests. This is still a function of individual interests and orientations. People who share common individual interests do sometimes combine to promote these interests in our politics. Indeed, our political system is set up to promote and entertain such a dynamic with the assumption being that all legitimate interests (defined as an aggregation of individual interests) should be represented in the policy arena. Thus we see many groups expressing varying and competing interests and policy preferences. Corporations, farmers, workers, businesspeople, environmentalists, those opposed to environmentalism, prolife advocates, prochoice advocates, rich people, poor people, and numerous other groups constitute organized interests relevant to politics and policy makers in the United States; this contributes to the representative character of government. Politicians in this mix are (ideally) regarded as brokers who make deals among various and competing groups that result in compromises that reflect as much as possible the preferences and interests of the widest variety of groups. That's the ideal description of American politics. This is the classic pluralist model.[15]

So American politics is about promoting interests. In American politics, when there is a debate between competing interests, there is a manufactured no-holds-barred contest to influence policy outcomes. There is absolutely nothing wrong with this in a free society. In fact, it is how our representative system is meant to work. The political contest waged often includes aggressive and effective dissemination of information that is less than accurate or truthful. The new articulation of this phenomenon is that the less-than-truthful or the inaccurate statements (i.e., lies) are "alternative facts." But, as we often say, that's politics. That's why Americans typically want to hear both sides of the debate. We assume that the competing sides in the political contest are equally flawed partisans or selective dispensers of fact and truth. The truth is often somewhere in between, or so the thought goes.

The assumption of the classic pluralist model is that interest group politics is actually representative of all citizens. That is a basic requirement if we are to regard such a system as democratic. Of course, careful empirical analysis conducted by social scientists over the past 60 years shows that only a small proportion of Americans are actually represented. Our policy-making process, responding to organized group pressures, leaves many Americans out. It has long been the case that some groups have much more influence than others, primarily due to the money and resources at their disposal that give them disproportionate amounts of influence over who gets elected and how they will govern.[16] Indeed, there are winners and losers in every policy debate. Public policy outcomes advance some interests, block or inhibit others, and result in an authoritative allocation of benefits and costs. The vast majority of Americans, attached to their microsocial existence and often oblivious to the macro-level battles in the policy arena, are confused by it all. They often do not feel represented, even by the groups that say they are representing their interests, and they are deeply suspicious and mistrustful of everything that goes on in the public or macro-level arena.

One of the unfortunate products of our politics is the degree of cynicism it produces. To the average citizen, the conflicts and battles played out in our politics are perhaps becoming too complicated to follow, especially given the casual and only occasional attention that most are willing to give it. The combatants themselves, the interest groups and the politicians and their supporters among the public, have increasingly divided into uncompromising ideological camps. Politicians' dependency on the big campaign contributions and other forms of support provided by powerful interest groups, together with the fear of well-financed opposition should they disappoint them, has made policy makers the captives of the wealthiest and most powerful interests. As these powerful interests battle each other for supremacy, the politicians beholden to them for support find it increasingly difficult to compromise. This has contributed in no small measure to the partisan gridlock we observe in the U.S. Congress today.

To the average citizen, bombarded by contentious misinformation from various interests and uncompromising posturing by politicians, and influenced by partisan news coverage on cable television and ideological trolls on social media, it seems that people are always saying things that are not true. This coincides with a declining respect for public officials, public institutions, and private institutions, and a near total mistrust of the media. The heavy focus on the game of politics, public opinion polls, and the drama of conflict, rather than the substance of policy, feeds directly into the erosion of public interest and trust. This feeds a general public cynicism based on the perception of a declining capacity or a lack of will among elected officials to solve public problems and a sense that the policies that do result are not representative of the average citizen. Such perceptions fuel a rapid acceleration in the decline of public confidence in all institutions.

If this description of politics in the United States is remotely accurate, the relationship between scientific expertise (i.e., knowledge) and political decision making is made all the more difficult. How can reliable scientific knowledge be made useful for policy making and for society at large in such an environment? How can we make scientifically and ethically sound decisions when the average citizen questions the democratic or representative legitimacy of the political process? How can scientific expertise in such an environment be used without compromising the quality and the reliability of knowledge? These important questions are relatively new, but especially given the politics of our age, they are important and are ever more widely discussed.

Until about the mid-20th century, scientific knowledge was considered dependable and unbiased. It was regarded as a source of objective truth by policy makers and most citizens alike.[17] Scientific experts advised the policy makers, their academic and professional credentials were the unquestioned foundation of their legitimacy in this role, and policy makers were reasonably comfortable with the legitimacy of the advice received. But since that time, this dynamic has changed considerably. The issues entering the political arena, combined with the intensification of ideological and partisan divisions over several decades, has created a greater distance between the values of policy makers and the work of science. Some political movements, such as the antinuclear movement or the environmental movement, injected new disagreements formerly outside of the science-policy nexus into the mix. These developments contributed to conflicts within the scientific community itself in some cases. In most cases, these developments led to increased conflict between the views of experts and the values of ideological policy makers, citizens, and laypersons.[18]

A 2015 Pew Research Center survey comparing the opinion differences between the public and scientists on a number of science-related issues showed that their level of disagreement is growing (see Table 1.3). Science still holds an esteemed place in the minds of most citizens, but that does not mean that they accept or agree with the findings of science. Yes, scientific

Table 1.3   Public versus Scientific Opinion

Issue	U.S. Adults	Scientists
Safe to eat genetically modified foods	37%	88%
Safe to eat foods grown with pesticides	28%	68%
Humans have evolved over time	65%	98%
Childhood vaccines should be required	68%	86%
Growing population is a serious problem	59%	82%
Favor offshore drilling	52%	32%
Climate change is due to human activity	50%	87%

Source: Pew Research Center Survey, http://www.pewinternet.org/2015/01/29
/public-and-scientists-views-on-science-and-society/

innovations are deeply embedded in American life. They have been connected to every aspect of our lives. Americans generally know and celebrate that. But recent polling suggests that they disagree, by a considerable margin in some cases, with scientific opinion on a number of issues. The PEW survey compared public responses with a representative sample of scientists connected to the American Association for the Advancement of Science on a host of issues and found some significant differences of opinion. It might not be surprising that average citizens tend to be selective in their acceptance of science and scientific conclusions.

We live in a time when the tensions between politics, citizens, and science are becoming more keenly observable. In a free and democratic society, this is not a bad thing. But it can bog down the already complicated process of policy making and prevent policy makers and society at large from responding to critical issues in a timely manner. Representative government must allow for public participation and vigorous debate. Yet the expertise of scientists must not be muted or rendered ineffectual in the political and policy-making processes. The twin specters of uninformed policies and too much technocracy animate many of the fears and concerns of scientists, policy makers, and citizens alike.

One study concerned with how scientists become involved in or influence policy making focused on U.S. policy debates over acid rain and climate change. It identified the influence of science on these issues in three different phases of the policy process—agenda setting, legislating, and implementation. This study suggested that scientific influences were most dominant in the agenda-setting phase. According to this analysis, this was the phase in which the prescriptions of scientists about a problem and what should be done about it had the greatest influence. It was in this phase that the agenda for future study and debate would likely become embodied in any subsequent legislation.[19] For example, scientific efforts to put acid rain on the legislative agenda resulted in the creation of the National Acid Precipitation Assessment Program, an institution through which research would continue and ultimately shape acid-rain policy in the later phases of policy making. Incredible as it may seem, some worry that such influence is undemocratic or technocratic and must be balanced by contributions from those demanding broader public engagement. Some have even attempted to develop a taxonomy of policy problems that distinguishes those dependent on scientists from those requiring broader public participation.[20] Much contemporary discussion and analysis is centered on what is seen as the problem of creating scientifically informed, yet democratic, policy.[21] The question appears to be whether giving scientists an important role in policy making leaves room for substantive contributions by diverse groups of nonscientists. This is a rather silly concern given what actually transpires in the policy-making process.

The concern about technocratic expertise eliminating broader participation in the policy process seems a bit exaggerated when one considers the immense involvement of the multiple groups attempting to influence public policy. Science rarely dominates or controls the dialogue, though it may influence some outcomes. But if you think about it honestly, science is rarely embraced when it presents us with new information that defines new problems that may require substantial alterations in how we live our lives or conduct our businesses. Most of the time, the warnings of science will meet fierce political and social resistance in such circumstances. There are certainly always those who, with a huge financial or political stake in preserving the status quo or ignoring the problem, will promote doubt and suspicion about the scientific advice of the experts. For example, in the not-too-distant past, science produced conclusive landmark studies documenting the dangers of DDT, tobacco, acid rain, and ozone depletion. In each of these cases, as is presently the case with climate change, economic and political interests worked very hard to present contrary perspectives and to delay or prevent policy action. It took a great deal of time to convince the policy makers and the public that policy action was required. Long political battles had to be fought and won before policy makers acted to address the dangers so accurately and completely documented by sound science.

The policy process is dominated by a combination of experts, special and well-funded interests, and partisan groups, and the average citizen is relatively disconnected and unrepresented in the process. But I would suggest that this is primarily due to how our political system operates, not due to any dangerous technocratic biases in our science. Indeed, science and scientists have been targeted as an adversary, even the enemy, too many times to suggest that it is dominating or influencing the policy process in harmful ways. The influence of ideology, partisanship, and special economic interests has more than diminished the ability of science to persuade. Politics, one might say, has brought science down to its level.

A bipartisan group of congress members, during the Obama administration, sponsored a bill that would have created a U.S. science laureate. The position would have been honorary and unpaid, thereby avoiding any debate or fuss over budgets and spending. The notion was that the science laureate would be tasked with getting U.S. schoolchildren excited about careers in science. The passage of this proposal seemed a sure thing as it had the support of leaders in both political parties. But the bill was dropped from the docket shortly after the American Conservative Union (ACU), an activist group, expressed concerns that a science laureate might express politicized views. President Obama, warned the ACU, could not be trusted. Why, he might name a liberal scientist to the post in an effort to sway public opinion on issues like climate change.[22] Even when they agree on something as simple as appointing a science laureate, our elected policy makers question the trustworthiness of scientists and insist that they are political. Even the popular concept of a scientist extolling the virtues of science and scientific careers to American schoolchildren is impossible to enact because of the mountains of doubt about science and scientists that have been manufactured in the political echo chambers. The things that dominate our attention and shape our perceptions serve only to remove us from reality, or so it often seems.

In the current political environment, it is easy to forget that most scientists are reluctant to engage in political discussion. They prefer to speak in the scientific realm and to be very meticulous in their work. Scientists are generally averse to rushing toward dramatic conclusions. Their inclination is to study matters in much greater detail and to demand much more painstaking independent verification before reaching dramatic conclusions. The more dramatic and potentially controversial the findings, the more painstaking and deliberate the process. If there is drama associated with the social, political, or economic implications of a scientific conclusion, scientists will be all the more deliberate and cautious to ensure that their conclusions will be understood and not dismissed for nonscientific reasons. The irony of this predisposition is that, as scientists refrain from political discussion, those who seek to create a climate of doubt regarding the legitimate and peer-reviewed conclusions of their work define the public discourse.

In the current political environment, it is easy to forget that throughout our history, public policy has been implemented on the basis of scientific knowledge that was neither absolutely certain nor 100 percent proven but that reflected the best consensus of expert scientific opinion. When the decision was made to ban DDT, the experts were not unanimous. But the prevailing scientific consensus, or the weight of the evidence as understood by the overwhelming majority of relevant scientific experts, supported the ban. Yes, some special interests opposed the banning of DDT. Even some scientists questioned the necessity for it. That is always the case. The point is, informed public policy was based on a consensus of relevant scientific experts. This consensus was accepted by policy makers and eventually by the public.

In a scientifically literate society, it is understood that even the experts do not always agree. Different conclusions may be reached even when they look at the same evidence because scientists may be focused on different dimensions of that evidence. A small number of earth scientists, for example, advocate earth expansion to explain continental separation, and they emphasize the limitations of plate tectonics theory. But plate tectonics remains the best theory for understanding the earth, according to most earth scientists. That's the scientific consensus, if you will. It is only very recently that policy makers have begun to demand 100 percent certainty or absolute proof as a precondition for policy action. This is the favored tactic of partisans who wish to deny or ignore what scientific experts recommend. It is always possible to find a view contrary to the consensus. Thus, "you have your science, and I have mine" often becomes the default position and political gridlock the more likely outcome.

Conclusion: A Troubled Relationship

At its most basic level, science is the pursuit of knowledge and its application to the natural and social world. It seeks the truth about the physical world, and it follows a systematic methodology based on strict rules of evidence. Politics, at its most basic level in the United States, is the process through which differing interests compete for advantage with respect to the formal allocation of values, benefits, and costs in our society. If you are a scientist, you may simply want to know stuff. You may not be particularly interested in politicians or worry about helping them to grapple with the things you have dedicated decades of your life to understanding that they are incapable of understanding. If you are a politician, you may be interested in science only insofar as it is useful to serving the communities or partisan interests you represent or supporting the decisions you have already made. But like it or not, scientists and politicians do need to collaborate. Whether combating new epidemic threats, developing environmentally friendly energy sources, creating new weapons systems, or combating climate disruption, the latest scientific and technical information is relevant and necessary in the policy process. Indeed, politicians may well need to become more science literate, and scientists may need to become more politically literate to create a relationship that produces the best informed policy decisions on technically complex issues.

We really must understand that scientists and politicians live in different worlds. Scientists work in a world where the result of their work is independent of social nuance. The aim is to remove context from the work. The results must be as broadly generalizable as possible. Politicians work in a world where context is everything. Politicians respond, first and foremost, to the world around them. Policy making is a response to partisan interests, the most pressing problems, and immediate concerns. The scientist deals with the grand questions of the universe. The politician deals with the minutia of the moment. The scientific approach to problems and issues leads to conclusions that are often at odds with religious and cultural beliefs. The political approach to problems and issues often leads to positions that do not agree with the facts. Scientists seek to understand and to educate. Politicians seek to win approval and to be reelected. For scientists, facts or data are the only things that matter. For politicians, facts and data are things to be massaged for decoration rather than information. Politicians routinely accept, dismiss, distort, or mischaracterize the opinions of experts to serve their ideology or appease their constituency. Many Americans are persuaded by partisan disagreements to dismiss scientists as being "elite," and the anti-intellectual strains of our culture quickly dismiss them as irrelevant or biased when it comes to matters of public policy. Thus it is that we hear stunningly ignorant talk of dinosaurs cavorting with humans (sorry but *The Flintstones* is not a historical documentary). We also see tales of climate scientists conspiring to create a global-warming hoax or see accusations that the health establishment is using vaccines to implement a socialist agenda and any number of other ridiculous claims competing with legitimate science for influence and attention in our public discourse.

It is not much discussed anymore, and many Americans may be unaware of it, but the Founding Fathers were science enthusiasts. Thomas Jefferson built the justification for the nation's independence on the thinking of Sir Isaac Newton, Francis Bacon, and John Locke. These men are regarded as among the creators of physics, inductive reasoning, and empiricism, respectively. Based on this foundation of science (i.e., the knowledge gained by systematic study and testing instead of by ideological assertion), our new form of democratic government was said to be self-evident. Contemporary Americans are not nearly so enthusiastic about science. They do not bat an eye when a new president in 2017 proclaims, contrary to all the evidence of science, "Nobody knows for sure if climate change is real."

Today we see all too frequently the denial of inconvenient science by partisans on both ends of the political spectrum. Science denial among Democrats is often motivated by unsupported suspicions or fears of hidden dangers to health or the environment. Examples include the belief that cell phones cause cancer or vaccines cause autism. Science denial among Republicans is motivated by an unreasonable antiregulatory fervor and religious fundamentalism. Examples include the belief that global warming is a hoax and the insistence that we should teach schoolchildren that there is a controversy about whether the planet was shaped by evolution over millions of years (as science has demonstrated) or by an intelligent designer over thousands of years. The Republican version of science denial is a bit more serious to the extent that it has been more frequently characterized by attacking the integrity of scientists and the validity of science itself as a basis for public policy when science disagrees with its ideology.

It is perhaps possible to best understand the relationship between politics and science in the United States in terms of four basic dynamics that are at work. The word "dynamics" here is understood as the motivating or driving forces in the policy process. In the context of our present discussion, there are four basic dynamics that influence the process. Depending on the issue being considered, one of these four dynamics may be the decisive factor. It is also possible that two or more dynamics are interacting and influencing policy outcomes. The truth of the matter is that none of these four dynamics really creates the best relationship between science and politics. One or more of these dynamics may from time to time lead to policy outcomes that serve the public interest in the short term, but none do that with consistency. Indeed, the inevitable collision of "scientific reality" with "political reality" most frequently limits the relationship in tragic ways.

The first dynamic is one that might be called the *collaborative dynamic*. This would apply to a situation where political policy makers and scientists are in basic agreement as to the policy goals and the scientific knowledge necessary for achieving them. The agreement is such that the level of trust between the two is unquestioned. Likewise, relevant special interests and/or the general public seem to be in agreement with the policy goals. An example in which this dynamic was especially strong would be the early U.S. space program, from the time of Sputnik to the Apollo moon landing. A perceived national crisis is often conducive to the collaborative dynamic. Whatever brings the collaborative dynamic about, it usually runs its course when it is no longer useful to the political purposes or policy goals of the policy maker.

The second dynamic, commonly observed in recent decades, is the *conflict dynamic*. Here we have an issue or concern where the conclusions or recommendations of science generate strong opposition from special interests and the political entities that they fund and support. Such opposition is based most frequently on economic or material interests. This opposition becomes a very influential and powerful part of the policy debate, often to the extent of muting or ignoring the science. The rise of environmentalism as a major concern in the 1960s is an example of this dynamic. Many corporate and private-sector interests spent huge sums of money trying to discredit the conclusions of environmental science and to impede the passage and implementation of policies that would impose regulations or restrictions that these interests felt harmed their bottom line. Those opposed to environmental policy initiatives argue that the earth is not as fragile as some environmentalists maintain. Their arguments focused on the alleged harms environmental policy posed for job creation, wage enhancement, and industry. Generally, chemical manufacturers, oil producers, mining companies, timber companies, real estate developers, nuclear power industries, and electric utilities have antienvironmental motives or agendas that are well represented in the policy debate. In the context of this dynamic, the ability of science to be useful to the development of public policy that truly serves the national interest is slow to evolve. It is the product of much conflict over an extended period of time. Too many years of conflict may delay this process such that the accumulation of avoidable damages accelerates and makes the corrective policy required both more complex and much more expensive.

A third dynamic at play, closely related to the conflict dynamic, might be called the *resistance dynamic*. Here we are referring to a situation where strongly held cultural values are in opposition to the findings or conclusions of science. Significant segments of the public may be opposed to science on religious grounds, for example, or for reasons of ideology. As history has shown us time and again, fundamental religious belief and science do not always see eye to eye. Biblical literalists often oppose the teachings of science where they introduce facts or findings that conflict with their views about supposedly moral issues. Evolution deniers come to mind here. The debate about the teaching of evolution and creationism is not one that generally gets decided by the science. The people who favor the teaching of intelligent design as an alternative to evolution are not motivated by science, nor are the politicians who support their preferences. Resistance to the science is the factor that is operative here, and the defense of beliefs or cultural values against the claims of science is the objective. The resistance dynamic is inevitable in cases where science, no matter how reliable its findings, runs toward conclusions that conflict with deeply held cultural values or biases. Like the conflict dynamic, the resistance dynamic may delay the process or render scientific advice powerless long enough to allow for the accumulation of avoidable damages.

The fourth dynamic is the unexpected or the *panic dynamic*. This might be a new pandemic threat or epidemic or a heretofore unknown disease. In such a situation, the science is working hard to catch up with events and to explain what is happening, the public is in a state of confusion or panic, and policy makers are without answers. Fear may grip the public, and politicians may be flummoxed as scientists seek to project a sense of calm and inform an uneasy public. New disease outbreaks are fairly rare and, when they do happen, they are often overstated. In recent years, there has been plenty of handwringing and worrying over bird flu, swine flu, and other diseases. Pandemic preparedness has been emphasized in some instances, and wisely so. And there have been recent outbreaks that have panicked the public and activated governmental agencies. In relation to the recent outbreak of the Zika virus, the Centers for Disease Control and Prevention issued travel guidance on affected countries, advised taking enhanced precautions, and provided guidelines for pregnant women, including considering postponing travel. Other governments or health agencies also issued similar travel warnings. This is a typical and appropriate response, but it overlooks perhaps the more important questions: How well prepared are we really for the new and unexpected? How well do we handle these developments? Might we be advised to do more preparation for the unexpected? Are we doing enough? Or are we just reacting after the fact? It can be argued that the weaknesses or flaws in the overall relationship between science and politics are most visible in the panic or crisis situations.

Summing up, public policy that seeks to incorporate or make use of science to reach informed decisions on technically complex matters is the product of many influences. In addition to the expert advice of scientists, special or vested interests may seek to influence outcomes to their advantage. Cultural values held by significant numbers of the public or conflicting values held by different publics may influence the process. Politicians, as the brokers seeking to bring about consensus or as partisans seeking to win the policy battle on behalf of one interest or one public versus another may intentionally muddy the waters of expert advice even further.

Depending on the issue and the context, the dynamic in the relationship between science and politics may be one of collaboration, conflict, resistance, or panic, or it may be a combination of two or more of these dynamics. Considering these dynamics, it is no surprise that *there is a chasm separating how science understands the world from how politics acts in the world.* Is it any wonder that we often seem to be confused and divided in our attempts to deal with the supremely difficult collective choices science and technology continually place before us?

Before we can reasonably hope to improve the relationship between politics and science, assuming that is possible, we need a better understanding of what that imperfect relationship has been up to this point. We need to better and more fully understand the dynamics that have defined that relationship and limited it. We need to examine the road we have traveled. Toward that end, five issues will be discussed as case studies that exemplify the four basic dynamics we have outlined. The birth of the U.S. space program offers a near-perfect example of the collaborative dynamic. The debates over climate change and hydraulic fracturing for natural gas provide two examples, in which the conflict dynamic may be the prevailing influence. The evolution-creationism controversy as it plays out in political and policy debate is a prime example of the resistance dynamic at work. Our fifth case, the examination of pandemic preparedness in the United States, will highlight the dynamic we have referred to as the unknown or panic dynamic. Again, more than one dynamic may be at work in each case. For example, it is almost impossible to say that there is any situation in which there will not be some conflict. It may be that cultural values are always a factor, and even in the midst of a conflict situation, we may see some collaborative efforts. But in each of the cases to be presented, one of the dynamics dominates or has the largest impact on the outcomes. Each case will demonstrate the inevitability of a collision between science and politics. This collision, it will be argued and demonstrated, however understandable and inevitable, is less tolerable than ever in our present age. We have reached a moment in human history when the cumulative impact of the collisions may become lethal. A new course must be charted.

As we use the five case studies to examine the imperfect relationship that has existed between politics and science, we may find that it is possible to understand where we fall short of the ideal and how we might approximate a closer fit to it. What is the ideal? In a perfect world, scientists would collect facts, politicians would develop policies based on those facts, legislators would pass laws to implement these policies, and government agencies would enforce the laws, most likely through regulations based on the same kind of facts. Because the laws, policies, and regulations would be based on the truth, they would work. Our problems would be solved efficiently, effectively, and economically. More subtly, we might say that science would give us our most reliable understanding of the natural world and, therefore, provide the best possible basis for public policy on subjects involving the natural world. Okay, maybe that is far too idealistic. But at least it might be possible to improve the relationship.

It should be possible for natural science to play a role by providing informed opinions about the plausible consequences of our actions (or inactions) and by monitoring the effects of our choices. Social science can do the same. It should be possible for policy makers to avoid making ill-informed decisions even as they represent different and competing interests. It should be possible to respect different values and yet factor the best scientific information into our thinking about public issues. It should be possible to improve scientists' communication to policy makers and the public. It should also be possible for scientists to better understand politicians and the public. As we complete the five cases to be examined, we will endeavor to make some concluding recommendations as to how these more limited but practical goals might be achievable. The premise of this book is that achieving these goals is the top priority for our present age. If at least that one point is driven home, the effort will have been worthwhile.

The great issues that will define our future, or determine if we will indeed have a future, are going to be decided in the relationship between science and politics. When it comes to big data, genetics, climate, energy, and virtually every area where science and politics must interact, there will be many interests pushing to spin the relevant science. It is not a foregone conclusion that we, as members of a democratic society, are ready and/or able to respond as informed citizens. If we do not understand science and the scientific method, or if we abandon rationality for ideology and/or misinformation manufactured to confuse and persuade us, how can we act with wisdom? As we elect our leaders, will we choose the best and the wisest proxies, and on what basis will they act? Most importantly, if our incomprehension and confusion about science and politics leads us to make the wrong choices, what will be the consequences? Will our inability to choose wisely compound our problems? Will we seek to empower a stronger hand to deal with the resulting mess, abandoning democracy in favor of expediency? For all the problems that we identify in the troubled relationship between science and politics, improving this relationship may be the most important challenge to be met if humanity is to thrive and survive.

There is no easy path to improving the relationship of science and politics. Yet, in a world whose politics will be increasingly dominated by issues that exist because of science or that cannot be recognized and managed without science, we must begin by recognizing the existence of the problems posed by the current troubled relationship and their gravity. The colliding worlds of science and politics must find a healthier way to interact. Our fate as a society and our future as a species may very well depend on getting this relationship on a more stable ground.

CHAPTER TWO

The Space Race: A Marriage of Necessity

As he looked into the October sky from his Texas ranch, Senator Lyndon B. Johnson strained to catch a glimpse of "that object which had been cast into the outer reaches of the world." For LBJ, it represented a "defeat as serious as Pearl Harbor."[1]

The launching of Sputnik 1 by the Soviet Union on October 4, 1957, rudely awakened Americans to a rapidly developing space age. Sputnik, the world's first artificial satellite, brought the Soviet Union into the technological spotlight and demonstrated that the country was capable of impressive new technological feats. It also sent a shock wave through the American public. Sputnik caused deep concerns among Americans who had felt a sense of technological superiority amid a postwar economic boom. Was the United States falling behind? Could Sputnik be a play on the part of the Soviets to put arms in space? Is space a necessary place to compete for world prestige?

Initial reactions to the Soviet accomplishment included shock, dismay, and alarm. Senator Henry Jackson called Sputnik a "devastating blow to U.S. scientific, industrial, and technological prestige." Senator Mike Mansfield called for a new Manhattan Project to regain missile superiority over the Soviet Union. Adlai Stevenson, Democratic presidential nominee in 1952 and 1956, expressed a concern shared by many when he said that "not just our pride but our security is at stake."[2] The American public was also quick to react. Six months prior to the launching of Sputnik, only one in five Americans could give a reasonably accurate description of a space satellite. Within weeks after Sputnik, 90 percent knew what a satellite was.

Indeed, people the world over demonstrated a keen awareness of and interest in the Soviet accomplishment. Worldwide reaction suggested a growing
respect for the Soviets.[3]

Americans typically expressed concerns about falling behind the Soviets technologically, about possible U.S. military vulnerability, about U.S.
scientific prowess, and about confidence in American leadership on the
world stage. Indeed, there was a sense of urgency in American public
opinion and a strongly felt need to respond to what was perceived to be a
crisis. For its part, the Eisenhower administration was neither shocked
nor alarmed. Defense Secretary Charles F. Wilson called Sputnik "a neat
scientific trick." President Eisenhower himself, initially concerned about
the possible military significance of Sputnik, was persuaded that there
was nothing that justified concern. The administration's first public
response concluded that there was no need to alter the U.S. research and
development program with regard to either space technology in general
or to ballistic missiles in particular. The president's immediate concern
was, in his words, to "find ways of affording perspective to our people and
so relieve the current wave of near hysteria."[4]

Generally speaking, the shock and anxiety felt by many Americans was
completely understandable. Sputnik launched in 1957, still very near the
beginning of the Cold War era. The Soviet Union, the feared enemy of the
United States in that conflict, had won the first part of the space race.
They had beaten the United States into space and, by inference, had
staked a claim to technological superiority over their adversary. The dangers this might portend were unthinkable to many citizens and to their
elected office holders.

The Eisenhower administration initially downplayed the satellite as a
"useless hunk of iron."[5] But to others, including a majority in the U.S.
Congress, the success of Sputnik seemed to herald a kind of technological
Pearl Harbor. Many in the United States and around the world saw Sputnik as an ominous leap ahead in prestige and military ability and speculated about whether the new missiles could actually hit a target with
nuclear weapons. President Eisenhower and some of his advisors, when
they did begin to realize the impact of the Soviet achievement on public
opinion, met to discuss what others perceived as alarming developments.
A memo of that meeting preserved the initial reactions of those present.[6]

It is interesting to see this memo of October 8, 1957, summarizing the
initial reactions of those inside the Eisenhower administration. A review
of its content reveals several interesting pieces of information. First, the
United States was taken by surprise by the Soviet launch. The Americans
had not anticipated that the Soviets would move so swiftly. Second, the

technological accomplishment was not threatening to those in the room, as U.S. capabilities were such that this feat posed no real threat to the United States.[7] Indeed, we could have launched our military rocket projects before the Soviets, but we had chosen not to rush a military launch ahead of a civilian or scientific launch. The United States wished to be perceived as being primarily interested in the scientific aspects of space technology. An American earth satellite must therefore be seen to proceed separately from military development, or so it was thought. There was no need for panic or alarm. Likewise, the president himself indicated that he saw no need for "a sudden shift in our approach."[8]

President Eisenhower, in part to address growing public concerns, did order an acceleration of the ballistic missile program. But even as he did so, he remained convinced that the international and domestic political and psychological significance of the Soviets' spectacular space achievement would be of short duration. Eisenhower was most emphatic in his belief that the United States should not become involved in a space race with the Soviet Union. But the public hysteria increased a month after Sputnik 1. The Soviet Union launched Sputnik 2 on November 3, 1957, this time with a passenger (a dog named Laika) aboard. Sputnik 2 weighed 1,120 pounds, and it orbited Earth for almost 200 days. This showed Sputnik was no fluke. American citizens were increasingly anxious to blast off and respond.[9]

Inevitably, and despite initial attempts to downplay the fears launched by Sputnik, a national sense of crisis evolved and created the conditions necessary for what would become a historic collaboration between the scientific and political worlds. It would be a collaboration driven by what was perceived as a political and national-security necessity. Understanding this historic collaboration sheds considerable light on the relationship between science and politics. The relative influence of the political goals over the scientific goals, the event-driven nature of the collaborative effort, and the inevitable limits of the collaborative dynamic in our political system demonstrate that science and politics can only temporarily and with limited effectiveness overcome their tendency to collide. If we examine the history of this unique collaborative moment, we will see that the conditions that facilitate collaboration are rare and that the natural impetus in the relationship between science and politics in the making of public policy is weighted more heavily toward the dynamics of conflict and resistance. In many ways, the space race was an aberration. It was a marriage of convenience doomed to be short-lived. To understand why science and politics collide so frequently and so foolishly in our present age, we must see how unusual and how limited this great collaboration was when all is said and done.

Research and Political Necessity: A Marriage of Convenience

Post–World War II rocket research in both the United States and the Soviet Union developed out of explicitly military programs designed to create ballistic missiles for the delivery of nuclear warheads. Both the United States and the Soviet Union realized the potential of rocketry as a military weapon and began a variety of experimental programs. Both the Americans and the Soviets had captured German World War II rockets in various stages of construction along with German plans for more advanced vehicles. Both utilized German scientists in the effort to develop their own military projects. At first, the United States began a program with high-altitude atmospheric sounding rockets. Later, a variety of medium- and long-range intercontinental ballistic missiles were developed. These efforts, originally motivated by military or national-security objectives, became the starting point of the U.S. space program. Missiles developed through military research and development efforts, such as the Redstone, Atlas, and Titan, would eventually launch astronauts into space.[10]

The United States proceeded with both military and scientific rocket research in the aftermath of World War II. The Eisenhower administration insisted that the military and scientific applications of space technology be maintained as separate entities. It was decided that the first U.S. excursion into outer space should be for peaceful or scientific purposes. The president had approved the launching of an American satellite in conjunction with U.S. participation in the International Geophysical Year activities that would span the period from July 1957 to December 1958. This project, called Vanguard, was independent of military research.[11]

The irony is that the U.S. Army could have orbited a satellite a year earlier than the Soviets did. In September 1956, Wernher von Braun and his team of scientists working in the army's Ballistic Missile Agency were ready to send a small payload into orbit in a test of the Jupiter C rocket. The Pentagon, in accordance with Eisenhower's wishes, directed that the fourth stage of the rocket be filled with sand rather than fuel and that no satellite be delivered into orbit. The United States would wait for its scientific project, the Vanguard, to proceed on its schedule and to be the first U.S. orbital insertion. Thus, the Soviets were able to get the jump on the Americans, so to speak. The U.S. Earth satellite program had begun in 1954 as a joint U.S. Army and U.S. Navy project. It was called Project Orbiter, and its objective was to put a scientific satellite into orbit during the International Geophysical Year. The initial plan was to use a military Redstone missile, but this was rejected in 1955 by the Eisenhower administration in favor of the navy's Project Vanguard, which used a booster advertised as more "civilian" in nature.[12]

On January 31, 1958, Explorer 1 was launched as part of U.S. participation in the International Geophysical Year. Almost four months after
Sputnik 1, the United States was in orbit. Explorer 1 was the first spacecraft to detect the Van Allen radiation belt. It returned data until its bat-
teries were exhausted after nearly four months. It actually remained in
orbit until 1970 and reentered the atmosphere over the Pacific Ocean on
March 31, 1970, after more than 58,000 orbits.[13]

The United States was in space, and it might be said that the race was
on. Yet President Eisenhower remained generally skeptical about space,
especially military uses of it. He was inclined to think that the scientific
value of space research and development was not commensurate to its costs
and that the military benefits were few, if any. Thus, he continued to
advance a cautious agenda. Eisenhower's criteria for deciding U.S. space
goals were spelled out in conservative terms. If a project were to be designed
solely for scientific purposes, its size and cost would have to be tailored to
the scientific job it was going to do. If a project were to have any ultimate
defense-related value, its urgency for this purpose would be judged in comparison with the probable value of competing projects. At least insofar as
President Eisenhower was concerned, there would be no space race. But
the U.S. Congress wanted an accelerated space program. It was ready and
eager to invest heavily in new rocket technology and in programs for science and education based on the premise that the Americans needed to
catch up with the Soviets. The public largely supported these initiatives.[14]

There was growing pressure on the president to reorganize the U.S.
space efforts. Congress was already moving on this. On November 25,
1957, in the immediate aftermath of Sputnik, Senator Lyndon B. Johnson
began hearings on American space and missile activities in the Preparedness Investigating Subcommittee of the Senate Armed Services Committee. This led on February 6, 1958, to the establishment of a Senate Special
Committee on Space and Aeronautics with the goal of establishing a new
space agency. On the House side of Congress, the Select Committee on
Astronautics and Space Exploration was created on March 5, 1958, and
was chaired by House majority leader John W. McCormack. There were
many questions to be answered. Should there be a new agency or one built
in an already established institution, such as the National Science Foundation, the Atomic Energy Commission, or the National Advisory Committee for Aeronautics? Or should it be part of a military agency? The army
and air force naturally were both keen for this last option based on their
missile work. The air force in particular felt that any manned spaceflight
program should be theirs alone. They argued that they did the flying, after
all. Should the new agency include aeronautical activities? Should it have
the power to implement international agreements? How should those

agreements be used as an instrument of foreign policy, and what should the new agency's relationship be with the State Department? The discussion was serious, and it was moving along very quickly.[15]

By the spring of 1958, the emerging legislative consensus favored the creation of a new civilian space agency. On April 2, President Eisenhower, in response to this consensus, sent draft legislation to Congress proposing the establishment of the National Aeronautics and Space Agency (NASA). After fairly quick and efficient congressional hearings during the spring of 1958, Congress passed the legislation, and President Eisenhower signed the National Aeronautics and Space Act into law on July 29, 1958. But even with the creation of NASA, largely in response to pressure from the Democratic-controlled Congress to provide some structure and coherence to U.S. space policy and implementation, the Eisenhower approach remained unchanged. Critics of the administration were shocked at its unwillingness to seek dominance in all aspects of space technology. Looking at the separate military and scientific projects, the lack of comprehensive and bold planning, and the small objectives of the Eisenhower administration, critics at the time talked of "space maze and missile mess."[16]

It is interesting to look at and put into context the objectives stated for NASA in the legislation enacted by Congress and signed by the president. The profound impact of Sputnik on the American psyche is revealed in these objectives (see Box 2.1).[17]

It would appear, looking at section 102 of the Space Act, that Congress was creating a technocratic institution. In other words, NASA would be an institution where technocracy is defined as the institutionalization of technological change for state purposes. This would lead to one of the largest state-funded and -managed R&D explosions in American history. NASA would thus appear to be aimed at sparking a technological revolution. In this sense, NASA can be seen as an engine of American international prestige. But NASA is also rooted in, or is a product of, the atomic diplomacy of its age. Some debates in Congress about the new agency were largely approached from within a framework of atomic energy, thereby influencing the shape of the new agency. Some of the objectives of the National Aeronautics and Space Act were clearly inspired by the Atomic Energy Acts of 1946 and 1954, especially the critically important relation of the Department of Defense to the new agency, the role of international cooperation, and the apportionment of intellectual property. Security issues great and small were involved in the founding of NASA.[18]

It was widely believed and feared that if the Soviets could launch a satellite into space, they probably could launch nuclear missiles capable of reaching U.S. shores. It was equally believed that U.S. leadership in

---

BOX 2.1. OBJECTIVES FOR NASA (SECTION 102 OF THE FINAL SPACE ACT)

1. The expansion of human knowledge of phenomena in the atmosphere and space;

2. The improvement of the usefulness, performance, speed, safety, and efficiency of aeronautical and space vehicles;

3. The development and operation of vehicles capable of carrying instruments, equipment, supplies, and living organisms through space;

4. The establishment of long-range studies of the potential benefits to be gained from, the opportunities for, and the problems involved in the utilization of aeronautical and space activities for peaceful and scientific purposes;

5. The preservation of the role of the United States as a leader in aeronautical and space science and technology and in the application thereof to the conduct of peaceful activities within and outside the atmosphere;

6. The making available to agencies directly concerned with national defense of discoveries that have military value or significance, and the furnishing by such agencies, to the civilian agency established to direct and control nonmilitary aeronautical and space activities, of information as to discoveries which have value or significance to that agency;

7. Cooperation by the United States with other nations and groups of nations in work done pursuant to this Act and in the peaceful application of the results thereof;

8. The most effective utilization of the scientific and engineering resources of the United States, with close cooperation among all interested agencies of the United States in order to avoid unnecessary duplication of effort, facilities and equipment.

Source: NASA, http://www.nasa.gov/exploration/whyweexplore/Why_We_29.html

---

technology was a necessity for national security. What is most clear is that, whether thinking in terms of military or scientific and technological advances, space was seen by a growing number of congressional leaders and citizens as a priority in a way heretofore not imagined. The investigations into the American space efforts that led to the establishment of NASA were premised on the assumed linkage between space exploration and vital national-security interests. As then Senator Lyndon B. Johnson saw it: "Control of space means control of the world."[19] He felt it was critical that the American people understand this and equally important that they provide the popular support necessary to sustain a long-term public enterprise that would be required for the United States to achieve and maintain scientific and technological superiority. This was essential as Johnson saw it because the masters of space would "have the power to control the earth's weather, to cause drought or flood, to divert the Gulf Stream."[20] More important to U.S. security and economic interests than having the ultimate weapon was having the ultimate position. As Johnson saw it, the ultimate position was the "position of total control over earth that lies somewhere in outer space." That position, said Johnson, must be our national goal. The United States must win and hold that ultimate position.[21] Its security and its future depended on that in the eyes of many. But despite the fact that Democrats in Congress and a growing majority in the public had a growing sense of urgency about all this, the Eisenhower administration did not share that feeling.

President Eisenhower wished to pursue incremental progress on military rocket research. This made sense to him as a practical national-security matter. But even with the creation of NASA, he was not aggressive about space exploration in general. NASA under Eisenhower would produce a 10-year plan for limited civilian space projects. The concept of pursuing preeminence in every aspect of space activity—civilian and military, scientific, and commercial, prestige-oriented and spectacular—was unthinkable. Eisenhower and his advisors considered such thinking to be extravagant. To them, the military and scientific contributions of such an effort did not seem substantial enough to justify such a national sacrifice of resources. Before he left office, NASA presented President Eisenhower with its plans for manned space exploration. He authorized only a limited manned space program. He saw no use for such a program beyond a very limited Mercury program.[22]

When presented with plans for a manned lunar project, something called Apollo, President Eisenhower was most emphatically negative. Apollo was an acronym used by NASA for "America's Program for Orbital and Lunar Landing Operations." Alarmed by the projected costs of such a program, he is alleged to have said, "I am not about to hock my jewels to send a man to the moon."[23] The general reaction at the meeting where the lunar proposal was presented is best described as one of bewilderment and amusement. As one presidential aide said of the request, "This won't satisfy everybody. When they finish this, they'll want to go to the planets." There was a good deal of laughter at that thought.[24] Clearly, the Eisenhower administration was not thinking that manned spaceflight was a priority. Indeed, it was not the major priority that defined the U.S. interest in rocket research in the immediate aftermath of World War II. Even in the late 1950s, manned spaceflight was an expensive and, many thought, unnecessary frill.

Let us take stock of where the U.S. space program was at the end of the 1950s. It is safe to say that it was military research that prompted the interest in rocketry. To some extent, U.S. efforts in this area were greatly advanced on September 20, 1945, the day that Wernher von Braun arrived at Fort Strong, a small military site on the northern tip of Boston Harbor's Long Island. It was the processing point for Project Paperclip, the government program under which hundreds of German scientists were brought into the United States at the end of World War II. Von Braun filled out his paperwork that day as the inventor of the Nazi V-2 rocket, a member of the Nazi Party, and a member of the SS. This SS membership linked him to those who were responsible for the deaths of thousands of concentration camp prisoners.[25] While his actions during World War II were surely supportive of monstrous deeds, he wasn't motivated by some inherent evil or personal belief in Nazi ideology. Von Braun was motivated by his childhood obsession with spaceflight and perhaps a somewhat uncritical patriotism. His main interest, his only interest really, was doing rocket research. For the United States in the aftermath of World War II, his talents as a rocket researcher were all that mattered.[26] For von Braun, the opportunity to continue his life's work was all that mattered.

So von Braun's main interest, above all else, was research in rocketry. His biggest dream was to land a man on the moon. All he really wanted to do was build rockets and launch them into space. After the war, he was attracted by the opportunities the United States promised and suspected that the U.S. military would support his continued research in rocketry.[27] Von Braun will perhaps always be viewed by some as a villain and by others as a visionary. Both might be true. He certainly exploited horrifying means to pursue his goals. But the constant focus of his life was his fixation on manned spaceflight. In many ways, his story mirrors the story of the U.S. space program. A marriage of convenience between military necessity, so defined by the policy elites, and rocket research fueled the early American efforts. The United States' space program was launched with the help of 118 German rocket scientists brought here from Hitler's Third Reich. It was born of continued research into the more efficient and deadly delivery of weapons of mass destruction.[28]

Research on rockets in the 1950s had a military purpose. The scientific exploration of space was not a motivating factor. The capacity to deliver weapons of mass destruction was the primary strategic objective to be served. This was the political goal. It is what the policy makers had already decided needed to be done. Scientists were a means to the end already agreed upon, not a means for defining new ends. Most scientists had favored the building of the original atomic bomb. Even as scientists were working on the original atomic bomb, which the United States dropped on Japan to end World War II, they were also looking into the possibility of an even more powerful explosive device, known as the "Super" or hydrogen bomb. Scientists were split on the development of the hydrogen bomb. Albert Einstein opposed its development. So did J. Robert Oppenheimer, who had led the Manhattan Project, which developed the A-bomb. Others who knew about the project thought it would be morally wrong to create a bomb that could destroy an entire city, or more, in a single blast.[29] The space race, as the policy makers saw it, was mostly about competing with the Soviet Union and achieving technological dominance in relation to national-security concerns. The science that supported the achievement of these objectives became a priority. But many of the scientists were forward thinking with respect to the scientific knowledge to be gained from space exploration. Some of them had their eyes on a different prize. The policy makers and the scientists, perceiving different realities and priorities in some cases, it might be said, nevertheless had a basis for collaboration that was mutually beneficial in spite of their differing motivations and priorities. Theirs was a marriage of convenience, but it was a marriage nonetheless. But as the 1950s closed out, the marriage had not yet been fully consummated. That was about to change.

Moonshot

In the late 1950s, Senator John F. Kennedy of Massachusetts was ready to vote to kill the manned space program. He agreed with President Eisenhower that its cost was out of proportion to its benefits. Even in the early months of his presidency, Kennedy seemed ready to dismantle NASA. Perhaps the only thing that delayed his movement on this was the enthusiasm of his vice president and former Texas senator, Lyndon B. Johnson, for the space program and for NASA, which was headquartered in Houston.[30] The president's thinking would change very dramatically on April 12, 1961. An early-morning phone call from press secretary Pierre Salinger rearranged the young president's thinking instantly. Salinger read an Associated Press bulletin to the president. It said, "The Soviet Union announced today that it had won the race to put a man in space."[31]

Russian cosmonaut Yuri Gagarin had just orbited the earth in a spacecraft called *Vostok*. In a matter of hours after the launch, President Kennedy was grilling his vice president and others: "Is there any place we can catch them? What can we do? Are we working twenty-four hours a day? Can we go to the moon before them? . . . If somebody can just tell me how to catch up! Let's just find somebody, anybody. I don't care if it's the janitor over there, if he knows how."[32] In March 1961, a mere three weeks before Gagarin's flight, President Kennedy had decided against funding the Apollo project. He had told NASA director James Webb that the American moon project was to be put on indefinite hold. Now, in the aftermath of Gagarin's accomplishment, space had been elevated in Kennedy's thinking to the level of a supremely important political contest that the United States could not afford to lose.[33] It is clear in hindsight that the president's major concerns were related to national security in the context of the Cold War. The fear that Soviet dominance of space would become a means for them to deliver attacks on the United States and her allies was the major concern.

On May 5, 1961, the president was meeting with his National Security Council in the Cabinet Room at the White House. The meeting, which was focused on several hot-button Cold War concerns, was interrupted by a new priority relating to our global competition with the Soviet Union. Captain Alan Shepard, a navy pilot, was about to become the first American in space. Kennedy led the National Security Council into an adjoining room to watch the launch. Shepard would not orbit the earth. He would be in space for about 15 minutes. His flight would arc 115 miles up and then land in the ocean some 300 miles from where it blasted off. The president had been told that the odds of a successful flight were set at 75 percent. The flight was, of course, a total success.[34] The president was happy for the success, but he knew bigger things had to follow. In his mind, winning the space race was a critical national-security priority.

On May 21, 1961, President Kennedy took the unusual step of delivering to Congress a second State of the Union address in just a few months. "While this has traditionally been interpreted as an annual affair," the president began, "this tradition has been broken in extraordinary times. These are extraordinary times."[35] National defense was his major focus. The president had asked for increased military spending in January, mostly for new missiles and warheads. In this special message to Congress just a few months later, he asked for an additional $2 billion in military spending. Most of this was for conventional weapons and military assistance to countries threatened by "wars of liberation." It was in the context of these defense-related concerns that the president announced, in the aftermath of Gagarin's orbital mission and Alan Shepard's suborbital flight, the American ultimate response to the Soviet challenge in space.

As President Kennedy now saw it, U.S. leadership and security required defeating the Soviets in the space race. "If we are to win the battle that is now going on around the world between freedom and tyranny, the dramatic achievements in space which occurred in recent weeks should have made it clear to us all, as did Sputnik in 1957, the impact of this adventure on the minds of men everywhere." Taking a dramatic pause, the president delivered the most dramatic pledge imaginable. "I believe this nation should commit itself to achieving the goal, before this decade is out, of landing a man on the moon and returning him safely to the earth. No single space project will be more impressive to mankind."[36]

One of the people President Kennedy had turned to for advice before tossing his hat over the moon was Wernher von Braun. "Can we beat the Russians?" Kennedy asked. Von Braun thought we had a "sporting chance" but only "with an all-out crash program." Von Braun thought we could land a man on the moon by 1968.[37] For Kennedy, winning in space was now the major political imperative. He told Congress that this would cost over $500 million in the coming fiscal year and as much as $9 billion altogether. By 1969, the price tag would actually be much higher.[38] The race to the moon was now a critical component in the U.S. Cold War strategy. Money would be no object in meeting this national goal.

It is hard to convey today the intense public interest in each launch during the early days of manned spaceflight. During December 1961 and January 1962, the Americans eagerly anticipated the first orbital flight by an American astronaut. Ten scheduled launches had to be cancelled due to bad weather or minor equipment failures. It seemed we would never orbit the earth. Finally, on February 20, 1962, John Glenn's *Friendship 7* launched from Cape Canaveral. Over 100 million Americans, including the president of the United States, were watching on television. Lieutenant Colonel Glenn reported his experiences and sensations back to mission control. All went well for a while. But there would be trouble. The small rockets designed to keep the capsule at a steady altitude began to misfire. The capsule began to bounce from side to side. Next, NASA ground controllers were concerned about an indicator light that suggested that the heat shield might be loose and/or breaking up. If the shield broke off, the capsule and the astronaut would burn up on reentry into the atmosphere. The mission was cut short. Colonel Glenn used manual controls to guide the capsule back into the atmosphere and down to a soft parachute landing in the Atlantic after three orbits of the earth.[39] When President Kennedy and Colonel Glenn met a few days later, the president admired Glenn's courage and wanted to know how it felt to be in space. Glenn wanted to talk about numbers and science.

The U.S. space program was now ready to go full throttle. President Kennedy had made it a priority for national security. He appealed to the spirit of adventure, to patriotic pride, and to what he called the cause of freedom. The United States responded with one of the greatest mobilizations of resources and work force in U.S. history. Between 1963 and 1965, the Soviet Union would launch the first woman and the first three-man crew and do the first spacewalk (i.e., extravehicular excursion). But the United States remained steadfast in its determination to meet Kennedy's challenge of landing on the moon before the decade was out. The Mercury program was designed to put humans in space and to learn about our capacity to function there. The Gemini missions, with two astronaut crews, were designed to develop space travel techniques to support Apollo's ultimate mission to land astronauts on the moon. All this was a national priority motivated by what was perceived to be a crisis. The only science that mattered was the science necessary to land a man on the moon. In other words, that mission already established by the political leadership of the nation dictated the primary role of science in this collaborative effort.

The driving force, the major motivation for racing to the moon, was of course military. It cannot be emphasized enough that the sense that the Americans needed to catch up or regain their edge over the Soviets dominated every decision. The United States had failed to appreciate fully the implications of long-range strategic rockets capable of delivering bombs to any point on the earth. The Soviet Union did, and it moved more swiftly forward in rocket technology. It took the United States years to overcome miscalculations of the late 1940s and early 1950s.[40] The United States had underestimated how fast missile technology would move. It had misjudged how fast new materials, miniature electronics, and better triggering devices would improve the yield of warheads and shrink the size of nuclear bombs. It had also failed to anticipate how quickly the Soviet Union would move. The United States viewed the missile as an artillery weapon. It regarded airplanes with atomic bombs as the main line of strategic defense. That would have to change. Intercontinental ballistic missiles quickly became important.[41] By the 1960s, these early miscalculations made a race to the moon necessary as a national-security priority. The United States needed to win the space race, and the race to the moon was perceived as the way to do so. This is what motivated President Kennedy's moon challenge.

The Cold War may have triggered the race to the moon, but scientists of course had long wished it were possible to go to the moon for some very different reasons. They were eager to study the early history of the solar system, perhaps the universe. The moon, compared to the earth, had remained virtually unchanged over billions of years. The idea of humans walking on the moon, stepping billions of years into the past as it were, floated the boats of the scientific community. For generations scientists had dreamed of reaching the earth's pale companion in the sky.

For many generations this dream seemed as though it would never be more than a fantasy. The Cold War and weapons of unbelievable mass destruction scared policy makers into action, but this fear would not have produced any results without the inspiration that had motivated a generation of rocket theorists and scientists to design the necessary instruments to set foot in the stars. The race to the moon, a military necessity and a political gamble for policy makers, was also an opportunity for visionary scientists. They were eager to embark on a great moon odyssey to reach upward in search of the possibilities of a new era.

It has been suggested that the drive to achieve the Kennedy challenge to reach the moon by the end of the decade might have waned if not for a sniper's bullet in Dallas, Texas, in November 1963. This suggests that the Kennedy assassination turned a political pledge into an unbreakable deadline on a national headstone aglow in the light of an eternal flame.[42] That may be just a bit of an overstatement, yet it is no doubt true that honoring the memory of President Kennedy may have played a role in keeping the dream alive and progress toward its achievement on target. But let us not forget that as the manned space program progressed through the Mercury and Gemini stages, the irresistible competition to end the moon's isolation of nearly five billion years was also a goal that propelled each step forward. Whether thinking militarily or scientifically, that was the only goal. In their 1994 book, Alan Shepard and Donald "Deke" Slayton chronicled the race to orbit the earth, to function in space, to walk in space, and to walk on the moon itself. From the perspective of men who had reached for and attained unprecedented heights, everything from the early manned *Mercury* flights through the *Gemini* flights was preparation for the main event. The moon was the objective. All else was a stepping-stone. But the moon was just the beginning, or so it was hoped by many, especially the scientists.[43] Such were the bold dreams of the age.

Apollo 1 was to be launched on February 21, 1967. Final prelaunch tests were taking place by the end of January 1967. On January 27, the three Apollo 1 astronauts (Gus Grissom, Ed White, and Roger Chafee) suited up and entered the Apollo 1 command module at 1:00 p.m. and hooked into the spacecraft's oxygen and communications systems. For the next five and a half hours, what is known as a "plugs out" test proceeded. This means that the umbilical power cords that usually supplied power were removed—the plugs were out—and the spacecraft switched over to battery power. The cabin was pressurized with 16.7 pounds per square inch of 100 percent oxygen. With everything just as it would be on February 21, the crew went through a full simulation of countdown and launch. There were communication problems as static made it impossible at times for the astronauts and mission control to hear each other. The astronauts impatiently began to wonder out loud how they were expected to get to the moon if they couldn't talk between a few buildings.[44]

At 6:31 on the evening of January 27, 1967, the Apollo 1 prelaunch test went from routine to tragic in a matter of seconds. One of the astronauts (probably Chaffee) reported, "Fire, I smell fire." Two seconds later White was heard to say, "Fire in the cockpit." The fire spread throughout the cabin in a matter of seconds. The last crew communication ended 17 seconds after the start of the fire, followed by loss of all telemetry.[45] The death of three astronauts on the launchpad was, up to that time, the worst tragedy NASA had suffered. The Apollo program was put on hold for a while. Naturally, an exhaustive investigation was made of the accident. It was concluded that the most likely cause was a spark from a short circuit in a bundle of wires that ran to the left and just in front of Grissom's seat. The large amount of flammable material in the cabin in the pure oxygen environment allowed the fire to start and spread quickly.[46] Corrections were made, and the program resumed its course with an even more intense dedication. The eventual success of the Apollo program is today considered by NASA as a tribute to Gus Grissom, Ed White, and Roger Chaffee and proof that their tragic loss was not in vain. Sad though it was, this tragedy was not allowed to scuttle the objective. Much like the sniper's bullet in Dallas, it served to reinforce the national determination to succeed.

Given the national-security drive that fueled the early space program and that gave birth to the race to the moon, it is often suggested that the scientific exploration of the moon was a secondary concern. Many of the engineers and perhaps even the policy makers are said to have thought that lunar science was a distraction from the basic task of "landing a man on the moon and returning him safely to the earth."[47] In fact, the community of lunar scientists was very small when President Kennedy declared the moon objective in May 1961. Be that as it may, in early 1962, NASA's Office of Manned Space Flight (OMSF) asked NASA's Office of Space Science to design and articulate an Apollo science program. NASA physicist Charles Sonett was appointed to head this effort.[48]

The Sonett group, consisting of 12 members and 9 consultants, included a U.S. Geological Survey geologist, a geophysicist, astronomers, and chemists. The final report produced by this group, published in December 1963, was the first in a series of Apollo planning documents. Its recommendations called for ambitious scientific exploration of the moon. They called for all-proposed Apollo landing sites to be photographed by an automated lunar orbiter prior to their final selection. These photographs would be used to make detailed lunar maps, thus saving precious time during landing missions. Astronauts would be able to begin geological fieldwork without first mapping their landing sites.[49] The group urged that each two-person landing team include a scientist-astronaut with a PhD in geology and significant field experience. The OMSF planning group envisioned lunar missions of five days of uninterrupted scientific exploration. The group also expressed the conviction that lunar exploration would continue beyond the Apollo program. Thus, with great optimism, they recommended a large number of landing sites to be explored over an extended time frame.[50]

NASA paid attention to the Sonnet report and considered all its scientific objectives. It paid attention to other lunar scientists as well. But the interplay of the multitude of technical, scientific, and political variables meant that scientists could not be given all they desired. They would not actually be given very much of what they wanted. In fact, only one scientist-astronaut ever reached the moon—geologist Harrison Schmitt on Apollo 17, the last manned lunar flight.[51] Schmitt was originally scheduled to fly on Apollo 18, but the likely cancellation of that mission led to great pressure from the scientific community to finally land an actual scientist on the moon. Harrison "Jack" Schmitt earned a PhD in geology from Harvard University in 1964. He had worked for the United States Geological Survey before beginning astronaut training in 1965.

In truth, it might be said that the Apollo program ended on July 21, 1969. That was the day when Apollo 11 astronauts Neil Armstrong and Buzz Aldrin stepped onto the moon. President Kennedy's mission had been completed, the challenge met. Additional flights followed, but Apollo 17 in 1972 would be the last. The last three of the scheduled Apollo flights (Apollo 18, 19, and 20) were cancelled for lack of funding. The U.S. Congress and the general public quickly lost interest in the moon after the success of the Apollo 11 mission. The Soviets did not present a continued interest in the moon. The race had been won and was clearly over. The scientific value of continued lunar exploration was not enough to keep the moon alive as a priority for the nation.[52] In fact, the scientific value of further lunar exploration seemed not to be appreciated outside of the scientific community.

Even for those of us old enough to remember the race to the moon and the early space program, it seems to have happened ages ago. Moon landings are ancient history. What was the race to the moon ultimately about? Space exploration and perhaps some science, as we have seen, formed only a very small part of the answer to that question. Apollo's importance was primarily ideological, national, social, and psychological. To comprehend the race to the moon, it is necessary to see it as part of a Cold War battle to demonstrate the superiority of capitalism over socialism. Beating the Soviets was the drive behind everything.[53] Science, as a means to that end, was a necessary and willing partner. The nature of the relationship, the collaborative dynamic that shaped it, was the product of a unique set of circumstances that are impossible to replicate. It is also impossible to sustain those conditions. Even if it is a crisis that leads to collaboration between science and politics, the crisis eventually ends, and when it does, the reset button ends the collaboration and, typically, the marriage.

The Collaborative Dynamic and "Big Science"

The period from the World War II Manhattan Project through the Apollo space program's moon landing in 1969 was a new and unprecedented era in scientific research. It was called an era of "big science."[54] The meaning here is that the scope of scientific research during and after World War II increased more than just incrementally. It exploded! Big science was distinguishable by the large number of individuals and research organizations involved in a project. The size of the budget, the size of the machinery and equipment needed to do the research, and the physical space necessary to carry out the research are also distinguishing characteristics. Big science was very different from the "little science" of the prewar era when a researcher worked in a small university or industrial lab on a project that might cost no more than a few tens of thousands of dollars. This landscape changed very dramatically.

The Apollo program was a prime example of "big science." The total cost of the Apollo program is estimated at $20 billion to $25 billion. This equaled more than a third of NASA's budget over the lifetime of the program.[55] As the Apollo program reached its peak, more than 400,000 people were involved in it. More than 20,000 institutions of higher education and industrial firms made contributions to it.[56] NASA employed large teams of scientists and engineers to manage its complex missions. It also encouraged innovations and inspired inventions far beyond its immediate scope of activity. Remarkable technological innovations of the age inspired by or invented by NASA served to advance progress in aeronautics research, space science, and space exploration. These innovations also had spin-off benefits that have greatly benefited humanity.[57] Thousands of technology spin-offs from the U.S. space program have improved national security, the economy, productivity, lifestyles, and much more. In fact, it is difficult to find an area of everyday life that hasn't been improved by these spin-offs.

NASA-developed technologies have benefited society in several broad categories. These include health and medicine, computer technology, transportation, public safety, environmental and agricultural resources, industrial productivity, and consumer goods. NASA's tangible impacts on our daily lives do not attract much attention perhaps, but such impacts were inevitable when you think about it. After all, NASA has always been engaged in very active and very broad research. Its basic goal is to explore space, of course, but to do that requires biological research, physical research, and so on. For example, NASA invented a system (a seven-step guide to monitor and test food production) to ensure that astronauts on their way to the moon would not get food poisoning. A quarter of a century later, the Food and Drug Administration and the Agriculture Department adopted that very same safety system for all of us. Within a year of adopting that system, the number of salmonella cases in the United States dropped by a factor of two.[58] In celebrating its 50th anniversary, NASA reflected on some of its positive societal impacts over the years. It is a very impressive record. Box 2.2 shows but a partial list of these impacts as compiled by NASA.

The transformative effects of the space program are seen all around us in numerous technologies and life-saving capabilities. We see them when weather satellites track hurricanes or provide us with information critical to our understanding of climate change. We see them when lives are saved through advanced breast-cancer screenings or when a heart defibrillator restores the proper rhythm of a patient's heart.[59] Of course, none of these impacts have anything to do with why we went into space to begin with. They have nothing to do with the sacrifices we were willing to make to reach the moon. And, impressive though these impacts definitely are, they have not been enough to attract our continued interest and investment in bold programs for space exploration. Indeed, science and scientific advancement was not the priority that propelled us into space. It was a necessary means to achieve another objective. Once that objective was achieved, science was not able to hold our interest. The marriage that had existed between our political objectives and science was doomed to run its course and to fade. Its many children (i.e., spin-offs) were not enough to keep the couple together, and their differences would inevitably become irreconcilable.

As we have seen, the Apollo program was the culmination of a massive program that was undertaken in the context of a competitive arms race between the United States and the Soviet Union. The Cold War saw significant military gains on both sides. This arms race eventually led to the development of rockets capable of reaching and striking enemy territory across the world. Soon each country was capitalizing on rocket technology to experiment with orbiting satellites and human spaceflight. The

Box 2.2. NASA's Positive Societal Impacts

1978: Teflon-coated fiberglass developed in the 1970s as a new fabric for astronaut spacesuits has been used as a permanent roofing material for buildings and stadiums worldwide. (By the way, contrary to urban myth, NASA did not invent Teflon.)

1982: Astronauts working on the lunar surface wore liquid-cooled garments under their space suits to protect them from temperatures approaching 250 degrees Fahrenheit. These garments, further developed and refined by NASA's Johnson Space Center, are among the agency's most widely used spinoffs, with adaptations for portable cooling systems for treatment of medical ailments such as burning limb syndrome, multiple sclerosis, spinal injuries and sports injuries.

1986: A joint National Bureau of Standards/NASA project directed at the Johnson Space Center resulted in a lightweight breathing system for firefighters. Now widely used in breathing apparatuses, the NASA technology is credited with significant reductions in inhalation injuries to the people who protect us.

1991: Tapping three separate NASA-developed technologies in the design and testing of its school bus chassis, a Chicago-based company was able to create a safer, more reliable, advanced chassis, which now has a large market share for this form of transportation.

1994: Relying on technologies created for servicing spacecraft, a Santa Barbara-based company developed a mechanical arm that allows surgeons to operate three instruments simultaneously, while performing laparoscopic surgery. In 2001, the first complete robotic surgical operation proved successful, when a team of doctors in New York removed the gallbladder of a woman in France using the Computer Motion equipment.

1995: Dr. Michael DeBakey of the Baylor College of Medicine teamed up with Johnson Space Center engineer David Saucier to develop an artificial heart pump – based on the design of NASA's space shuttle main engine fuel pumps – that supplements the heart's pumping capacity in the left ventricle. Later, a team at Ames Research Center modeled the blood flow, and improved the design to avoid harm to blood cells. The DeBakey Left Ventricular Assist Device (LVAD) can maintain the heart in a stable condition in patients requiring a transplant until a donor is found, which can range from one month to a year. Sometimes, permanent implantation of the LVAD can negate the need for a transplant. Bernard Rosenbaum, a Johnson Space Center propulsion engineer who worked with the DeBakey-Saucier group said, "I came to NASA in the early 1960s as we worked to land men on the moon, and I never dreamed I would also become part of an effort that could help people's lives. We were energized and excited to do whatever it took to make it work."

2000: NASA's "Software of the Year" award went to Internet-based Global Differential GPS (IGDG), a C-language package that provides an end-to-end system capability for GPS-based real-time positioning and orbit determination. Developed at NASA's Jet Propulsion Laboratory, the software is being used to operate and control real-time GPS data streaming from NASA's Global GPS Network. The Federal Aviation Administration has adopted the software's use into the Wide Area Augmentation System program that provides pilots in U.S. airspace with real-time, meter-level accurate knowledge of their positions.

2000: Three Small Business Innovation Research contracts with NASA's Langley Research Center resulted in a new, low cost ballistic parachute system that lowers an entire aircraft to the ground in the event of an emergency. These parachutes, now in use for civilian and military aircraft, can provide a safe landing for pilots and passengers in the event of engine failure, midair collision, pilot disorientation or incapacitation, unrecovered spin, extreme icing and fuel exhaustion. To date, the parachute system is credited with saving more than 200 lives.

2005: Two NASA Kennedy Space Center scientists and three faculty members from the University of Central Florida teamed up to develop NASA's Government and Commercial Invention of the Year for 2005, the Emulsified Zero-Valent Iron (EZVI) Technology. Designed to address the need to clean up the ground of the historic Launch Complex 34 at KSC that was polluted with chlorinated solvents used to clean Apollo rocket parts, the EZVI technology provides a cost-effective and efficient cleanup solution to underground pollution that poses a contamination threat to fresh water sources in the area. This technology has potential use for the cleanup of environmental contamination at thousands of Department of Energy, Department of Defense, NASA and private industry facilities throughout the country.

Source: NASA website, http://www.nasa.gov/50th/50th_magazine/benefits.html

arms race jumped from the atmosphere to low earth orbit and ultimately to the moon. This was the ultimate high ground in the arms race. Space became the field for what was seen as a necessary and dramatic public demonstration of U.S. military and technological might.[60] This perceived necessity was the driving force behind the effort and the unique condition that created the possibility for a collaborative dynamic to shape the relationship between politics and science during the Apollo era.

As we noted in chapter 1, "collaborative dynamic" refers to a situation where political policy makers and scientists are in basic agreement as to the policy goals and the scientific knowledge necessary for achieving them. The agreement is such that the level of trust between the two is unquestioned. Likewise, relevant special interests and/or the general public seem to be in agreement with the policy goals. It should be emphasized that this situation is extremely rare. The hot environment of the Cold War led to the commitment of tremendous amounts of political capital and governmental spending in support of a first-strike nuclear infrastructure. A significant portion of this spilled over to the scientific and aeronautical fields, which were associated with a peaceful and more optimistic message. But that happy by-product was not the driving force behind the collaborative dynamic.

The driving force that created a collaborative dynamic between politics and science, at least through the Apollo 11 moon mission, was a perceived national crisis that animated the nation across all partisan divides. This is a rare thing in American history, usually associated with war or a total economic collapse. The crisis must be so severe and recognition of it must be so universal as to convey the sense that business as usual is an unacceptable alternative. The crisis must be seen to require separating it from all other issues and problems. It must be prioritized for action above all else because of its centrality to our national and/or international interests. Manned spaceflight, and the moon mission, did not happen in a political vacuum. Space exploration, as we have seen, became a priority and a widely perceived necessity to demonstrate the military and technological might of the United States. It was a spin-off of the drive to develop rockets and vehicles that could travel higher and faster than their Soviet counterparts. The space program happened alongside of increasing U.S./Soviet tensions. A series of geopolitical crises, culminating perhaps in the Cuban missile crisis, added fuel to the fire.

The hot environment of the Cold War created the conditions that allowed for, demanded in the eyes of policy makers, an incredible and unprecedented commitment of political capital and governmental spending. Scientists and engineers may have had peaceful and optimistic motives that excited them in their work, but their work was supported in the political and policy process for entirely different reasons. The primary objective might be best described as the creation of a first-strike infrastructure and the deterrence of the Soviet adversary. The pursuit of this objective trickled down into the scientific and aeronautical fields. The pursuit of this objective also created a different way of thinking about policy making generally.

In most policy debates, policy makers will think and act in an incremental fashion. The typical mode of decision making is incremental, which considers only a few options for addressing issues or problems. These options most frequently differ only marginally from existing policies. Given the time and costs associated with a more comprehensive assessment and new policy initiatives, and the considerable and intense divisions among the special interests and the policy makers that work against dramatic change, incremental and negotiated adjustments in existing policy are a preferred shortcut to the resolution of the policy problem. As a practical matter, it is easier and less politically risky for the policy maker to build consensus around small adjustments or tweaks in existing policy than to promote new policy options that deviate in a significant way from existing ones. In this context, the more limited and incremental approach of the Eisenhower administration to the space program is the more or less normal course in the American policy process. Even John F. Kennedy, as a senator and later in the first months of his presidency, followed that more limited path. But the unfolding events of the Cold War created the rare opportunity for a more dramatic pace and a more comprehensive approach to addressing a national policy problem. It provided that rare spark that elevated the problem and ignited a crisis. It broke the incremental mold and created the conditions for something radically new and different. It also created new opportunities for rare but more robust partnerships and collaborations between politics and science.

When President Kennedy articulated the goal of "landing a man on the moon and returning him safely to the earth" in 1961, before the decade was out no less, it was a monumental goal. As we have seen, the technological challenges alone were significant, and the effort that would be required to meet the goal was both unprecedented and expensive. But a remarkable national effort was soon under way. Why was this effort undertaken? Why did the American people support the effort? The crisis at hand, the Cold War, was the driving force. Americans across party lines accepted that American technological superiority, including superiority in the space race, was a necessity in relation to our most pressing national-security interests. The cost of doing all of this was very, very great. But the option of not doing it was simply unthinkable to both the policy makers and to the public. That is a rare dynamic in the political realm. The Apollo program would never have been so eagerly and efficiently pursued absent the crisis.

It can be argued that the space program peaked in 1966. At this time, the final Gemini mission had been completed and the final push toward the Apollo project was under way. In 1966, NASA received its highest budget ever (at 4.5 percent of the total U.S. federal budget or $5.9 billion). But even before Apollo 11 completed the first successful moon landing just three years later, the social and political infrastructure and support for the space program had begun to wane.[61] A total of six Apollo moon landings occurred between 1969 and 1972. But the priorities began to change immediately after the first moon landing in July 1969. NASA began to revive plans for a space station. By 1970 it was announced that Apollo 20 would be cancelled in favor of the pursuit of a new venture: Skylab. Indeed, by the fall of 1970, NASA announced that Apollo 17 would actually be the last Apollo moon mission. In 1971, the Nixon White House had planned to cancel the Apollo program after Apollo 15. In the end, two additional missions were kept in place.[62]

The shift in NASA's priorities in the 1970s reflected the lack of will-power by policy makers and the rapid decline of public support for new exploratory missions to the moon and the planets. As NASA focused on the space shuttle and low-orbit missions, the physical infrastructure necessary for lunar missions vanished fairly quickly. The technical and manufacturing apparatus supporting both military and civilian operations began to wind down. With the emerging détente with the Soviets and the Strategic Arms Limitations Talks (as well as subsequent negotiations that would freeze the numbers of missiles the United States and the Soviet Union could deploy) the world was altered drastically. The urgency that had fueled the Cold War crisis no longer existed. The Soviets showed absolutely no interest in the moon. The arms race cooled, and as it did, it shrunk the public and political support for efforts to expand our exploration of the moon and of deep space.

On December 14, 1972, the last human to walk on the moon took humankind's last lunar steps at the conclusion of the Apollo 17 mission. In the decades that have passed since those last steps, our national interest in returning to the moon has been close to nil. Gene Cernan, the last human to step on the moon, expressed as his final thought before leaving the lunar surface the belief that we would, after a time, return to the moon. He concluded his final statement saying, "We leave as we came and, God willing, as we shall return, with peace and hope for all mankind." We have yet to return. By the 1970s, the levels of federal spending that NASA had received during the decade of the 1960s was considered untenable by the public and by policy makers. The nation had grown financially wary, and the nation's priorities had shifted dramatically. There would still be room for space spending, but under very tight constraints. NASA would be limited in the coming decades to more modest research and scientific missions, including things like the 1973 Skylab program, the space shuttle program, and a variety of robotic probes and satellites. U.S. operations were refocused on low-earth orbit activities. There have also been admirable international programs, such as the International Space Station, and major scientific missions, such as *Mars Pathfinder*, *Opportunity/Spirit*, and *Curiosity*. But these activities, important as they may be, have never quite captured the public imagination or the level of political and public support for space exploration experienced in the 1960s.

The space program in the aftermath of Apollo quickly regressed into the mode of a boring incremental normality. It regressed from being an important political priority to a normal or incremental mode in the policy process. The space program never again soared as high as it did in the decade of the 1960s. Once the crowned jewel of the American policy arena, it is now more like a forgotten stepchild. The collaborative dynamic that existed for the race to the moon could not survive the end of the crisis that had promoted it. The fact is, the partners were never perfectly compatible. The fear that motivated the politics was conveniently married to the ambition of scientists who wanted to create rockets, land a man on the moon, and explore the far reaches of space. The result of the merger between political fear and scientific ambition galvanized the nation and its politics for but a brief time and led to a collaboration that brought humankind to greater heights than were previously imaginable. Once the fear dissipated, the ambitions of the scientific partners were no longer relevant to the politicians or to the public they served. Indeed, as the fear was no longer a factor, it has been said that the political and public romanticism of that brief moment in history has been replaced by an overwhelmingly apathetic attitude toward space. Where there is apathy, most relationships will struggle. They usually end. Some suggest that the rise of China's space program may change this apathetic attitude toward space in the future as fear surges once again. This, some say, may rekindle the fire that lit the collaborative candle that launched the space age. But in all honesty, one must strain to see even a hint of this rekindling.

Conclusion

U.S. space exploration, as we have seen, has always been tied to the national security and political aspirations of nation-states. Space enthusiasts should never forget the dominant role politics played in shaping the contours and the outcomes of the space race. As much as some of us may wish it to be otherwise, scientific exploration and the advancement of knowledge were not the primary motivations behind the space race. They were surely important goals, but they were not the reason we entered space. They were not the reason we went to the moon. It was the politics of national security and the perception of a national-security crisis that dictated the terms and defined the relationship between politics and science. *The dynamic was collaborative only because the science was needed and was able to provide the expertise necessary to achieve what the politicians had already decided they wanted to do in response to the crisis.* This is the only possible formula for a truly collaborative relationship between politics and science.

Let us briefly restate what we have already said regarding the post-Apollo era. Within a year after Apollo 11 had landed and met the Kennedy challenge, politicians' priorities changed. The urgency that had fueled both the Cold War arms race and the space race evaporated at a fairly rapid pace. As the 1970s began, an emerging détente between the Americans and the Soviets cooled things even more rapidly. The Strategic Arms Limitations Talks had begun to freeze the number of nuclear missiles that could be deployed by the Americans and the Soviets. Political and public support for the efforts and costs associated with the space program began to wane. NASA itself, in light of this changing political environment, had to reprioritize. In 1971, less than two years after the first man walked on the moon, the Nixon administration announced that it would completely cancel the Apollo program after the flight of Apollo 15. Ultimately, two more flights were kept on the schedule, but Apollo and the moon were over and out. NASA would focus on a new set of priorities, the Skylab project and later the space shuttle.[63]

The physical infrastructure that had supported the lunar missions vanished. Saturn V rockets were no longer manufactured. The entire technical and manufacturing apparatus that had supported both civilian and military operations in space began to wind down. This is not to say that American space exploration ceased or did not remain a significant policy priority; it is only to suggest that the urgency and support for this priority dropped off the charts compared to the 1960s Cold War era. Until recently, the more limited objectives of the space program were focused entirely on low-earth orbit activities. There have been admirable collaborative international programs. As previously mentioned, these included such things as the International Space Station and major scientific efforts such as *Mars Pathfinder, Opportunity/Spirit,* and *Curiosity.*[64]

Interestingly, a Harris survey taken in 1970—less than a year after the first moon landing—showed that a majority of Americans (56 percent) thought the moon landing was *not worth the money spent.* Over the past 40 years, American citizens when polled have consistently said that the United States spends too much on space exploration. At no time over these four decades has more than 22 percent of the public said that the United States spends too little on space exploration.[65] But it is perhaps encouraging to note that the unwillingness of Americans to spend tax dollars on space exploration has not dimmed their expectations and optimism about the future of it. The American people do not want to pay for it, but this doesn't mean Americans aren't interested in exploring the possibilities of space. In fact, Americans generally see space exploration as a very positive thing. A Pew Research Center/*Smithsonian* magazine survey found that a third of Americans said they believe there will be manned long-term colonies on other planets by the year 2064 despite evidence suggesting the difficulties of accomplishing that. Also, 63 percent of respondents to a 2010 survey said that they believe astronauts will have landed on Mars by 2050. More than half said that ordinary humans will be able to participate in space travel.[66] Finally, NASA is held in high esteem by the American public. About three-quarters of Americans view NASA favorably—second only to the Centers for Disease Control and Prevention among federal agencies—according to a 2013 Pew Survey.[67]

It seems contradictory that Americans have such a positive attitude about space exploration but are unwilling to pay for it or think it is too costly. The Pew studies show that Americans strongly preferred increased spending on programs closer to home, including education (76 percent), public health (59 percent), and developing alternative energy sources (59 percent).[68] The policy makers by and large agree with these public preferences. This does not mean that the United States will not continue to make progress in scientific space exploration, but it may be pursued in a very different way than it was in the past. Newer private-sector players in the space field (e.g., Space X and Orbital Sciences Corporation) may lead to a new generation of infrastructure being developed and constructed. Private players may make space tourism possible in upcoming years. Some believe that space for business purposes may lead to initiatives like space mining, space colonies, and so on, for profit. These are just a few of the dreams or fantasies that occupy the mind. But how much of this space business is motivated by science? Probably much less than we'd like to admit.

It is arguably true that the best reasons for exploring the moon and the planets and bodies in our solar system are purely scientific. Such exploration could contribute to the greatest endeavors within the human capacity to achieve. They could allow us to understand the creation of our planet and solar system. Such exploration could lead to scientific and technological advances that enable us to address the issues of greatest concern in relation to the health and survival of our planet. But as we should see quite clearly as we reflect on the Apollo era and its aftermath, the scientific promise will not spark the collaborative relation between science and politics that is ultimately ideal for the achievement of these exalted goals. It would appear that only the urgency of a crisis at our doorstep that threatens the security, safety, or survival of the nation can pull politics and science together in a genuine collaborative effort. In most other times, noncrisis and/or relatively stable times, the relationship will be very different.

Most of the time, the dynamic that will apply to the relationship between politics and science will be a conflict dynamic, and competing economic and political objectives and interests will, as they play themselves out in the political arena, impede purely scientific objectives. Even in the best of times, there are always those who say we cannot afford to invest public resources or take public policy initiatives in science. The focus on immediate needs or concerns is frequently used to argue that support for research is somehow a luxury we cannot afford. At moments defined by the perceived economic or political necessities of the day, science may even be seen to pose a threat or inconvenience that must be met with resistance. Finally, because competing interests have much at stake, they will support the science that furthers their competitive interests. We next turn our attention to this dynamic. "You have your science, and I have mine" is the order of the day in the political and social arena, where the conflict dynamic defines relationships.

CHAPTER THREE

Climate Change: A Classic Case of Conflict

It is often said that the contemporary American environmental movement began on June 16, 1962. It was on that date that *New Yorker* magazine published the first of three excerpts from Rachel Carson's new book, *Silent Spring*. Controversy ignited immediately. This landmark book documented how DDT and other pesticides had irrevocably harmed animals and contaminated the world's food supply. The book's most alarming chapter depicted a nameless city of the future in which all life—from fish to birds to apple blossoms to human children—would be destroyed by the insidious effects of DDT.[1] The book, which received remarkable attention upon its publication, alarmed readers and ignited new concerns over environmental issues. It also generated a piercing howl of indignation and intense opposition from the chemical industry.

An executive of the American Cyanamid Company responded to *Silent Spring* by saying, "If man were to faithfully follow the teachings of Miss Carson we would return to the Dark Ages, and the insects and diseases and vermin would once again inherit the earth."[2] Monsanto went so far as to publish and distribute 5,000 copies of a brochure parodying *Silent Spring*. Entitled "The Desolate Year," this brochure emphasized the devastation and inconvenience of a world where famine, disease, and insects ran amok because chemical pesticides had been banned.[3] Many of the chemical industry's attacks against this book were more personal than scientific. Some went so far as to question Carson's integrity and even her sanity.

Rachel Carson's passionate concern in *Silent Spring* was for the future of the planet Earth and all life on it. She called for humans to act thoughtfully, carefully, and as responsible stewards of the living earth. She suggested a needed change in how democracies operated in relation to new and expanding environmental concerns. She felt it important for individuals and groups to question what poisons their governments allowed others to put into the environment. She identified human hubris and financial self-interest as the crux of the problem and had the audacity to ask if we could master ourselves and our appetites to live as an equal part of the earth's systems and not perceive ourselves as the master of them.

The intense criticism she received after the publication of *Silent Spring* was such that she spent the rest of her life (she died in 1964) defending herself. Most of these criticisms were from the industrial interests that manufactured the pesticides and chemicals she warned about. Some were also from scientists who were genuinely convinced that the risks posed by these chemicals were not excessive. But history, and science, would show that her concerns were justified. In 1963, in a review in the London *Observer*, W. H. Thorpe of the Department of Zoology at Cambridge University defended Carson's work. He said that those in the scientific community who disagreed with Carson "make a great mistake because most of them are narrowly trained chemists that have no knowledge or comprehension of the incredibly complex interrelations and interactions of living things."[4] The conflict generated by the publication of *Silent Spring* is an example of the ever more common by-product of the increasingly troubled relationship between science, politics, and economics in our contemporary society.

Ever since Henry David Thoreau spent two years, two months, and two days in second-growth forest around the shores of Walden Pond (on land owned by Ralph Waldo Emerson), the critique that much of the western world is a consumerist force separated from and destructive of nature has been a theme of some considerable interest.[5] Thoreau's *Walden or Life in the Woods* was published in 1854, but it would be over a century before this critique and environmentalism became one of the defining issues that animated our partisan political discourse. American environmentalism wasn't born in the 1960s, of course. It wasn't created by Rachel Carson either. As early as 1690, colonial governor William Penn required Pennsylvania settlers to preserve one acre of trees for every five acres cleared. In the 1760s, Benjamin Franklin led a Philadelphia effort that attempted to regulate waste disposal and water pollution.[6] President Theodore Roosevelt is revered as a lover of nature, and the Sierra Club existed long before the environmental movement of the 1960s. What did change with the 1960s was that the ethic of environmental conservation spread to the masses as never before. This created a sense of hope and possibility.

People believed that environmental protection mattered and that they could bring about a positive change. Indeed, let us not forget that the end of the decade and the early 1970s brought about the creation of the Environmental Protection Agency and the passage of the Clean Air Act. Many credited Rachel Carson for igniting the contemporary environmental movement by writing the right book at the right time,[7] but the introduction of environmentalism to the political and policy agendas of the 1960s and 1970s also introduced new and more intense conflicts between science and politics. These conflicts have frequently worked against the best interests of the public. Unfortunately, they have most often succeeded at defending the financial and material interests of corporate and economic interests at the expense of the broader public interest.

It is true that much of what we might call the environmental agenda has some bipartisan support. It even has corporate support, or so it is made to seem when we hear talk of corporate citizenship and corporate concern for the environment. Be that as it may, the environmental agenda meets fierce resistance from the corporate and private sector for some very obvious reasons. An ever-expanding growth-oriented capitalism and a fragile finite ecology are, by definition, on an unavoidable collision course. The process of capital accumulation often treats the planet's life-sustaining resources (arable land, groundwater, wetlands, forests, fisheries, ocean beds, rivers, air quality) as dispensable ingredients of limitless supply to be consumed or rendered toxic at will in the name of growth and expanding profit margins. Environmentalists warn that the support systems of the entire ecosphere—the earth's thin skin of fresh air, water, and topsoil—are at risk. They are seriously threatened by scientifically documented things such as global warming, massive erosion, and ozone depletion. Environmentalists argue that capitalism, if unregulated, poses inevitable threats to our ecology. The corporate or private sector answer to the environmentalist argues that regulatory schemes or public policies that are meant to protect the environment will threaten economic growth, result in a loss of jobs, destroy the economy, and negatively impact the quality of life. They also cast doubt about the legitimacy of the science and provide their own "science" (or "alternative facts") to show that the earth is not as vulnerable as the environmentalists say it is. In short, they suggest that all advocates of the environmental argument are motivated by some dangerous political ideology rather than legitimate science.

As noted, the successes of the environmental movement in the 1960s and 1970s resulted in a growing public awareness about the need (and in positive steps taken) to protect the environment. Even some of the corporate and private-sector opponents to environmental policy responded with a show of support. Much of this support amounts to what has been called "greenwashing." Greenwashing (a compound word modeled on "white-wash") is a form of corporate spin in which green PR or green marketing is deceptively used to promote the mostly false perception that an organization's products, aims, and policies are environmentally friendly. This is good PR and good politics, but where the bottom line is concerned, the inherent tension between economic and environmental concerns *always* leads to conflict. This conflict is so intense that it frequently overshadows the scientific realities of the issues at stake and distorts public perceptions to the point that science cannot be heard over the loud and ever-present political theater that plays itself out in our public discourse.

What we are calling the conflict dynamic is frequently the defining characteristic of issues where public policy and science must intersect. The conflict dynamic, as we noted in chapter 1, is one in which we have an issue or concern where the conclusions or recommendations of science generate strong opposition from special interests and the political entities that they fund and support. Such opposition is based most frequently on economic or material interests and becomes a dominant part of the policy debate, often to the extent of muting or ignoring the actual science. The rise of environmentalism as a major issue in the 1960s is an example of this dynamic. And it is very much a driving force in our national discussion about an issue that scientists are presently telling us may be the most important challenge facing humanity. It is a challenge, they say, that will become more intense as it impacts the life of every living creature on the planet. Despite the solid conclusions of science, climate change remains a contentious issue that divides the nation. It is a classic case of the conflict dynamic at work.

The doubts, disagreements, and confusion of the American public and political decision makers about climate change are all too commonly observed. They are really most peculiar (insane, actually) in the context of the past 150 years of scientific research and study of climate change. It is disturbingly accurate to say that our public policy and our public discourse have both lagged very far behind the science. Indeed, as the science has become more settled, it seems that the policy and political components of the climate discussion have become more unsettled and contentious. This has been, we shall see, a matter of design, as many participants in the public dialogue have sought to create doubt, sow seeds of confusion, misdirect the public, and profit by the delay brought about by disagreement.

Climate Science

In June of 1988, a hot June at that, an important moment had arrived. Dr. James E. Hansen, director of NASA's Institute for Space Studies, testified before the U.S. Senate Energy and Natural Resources Committee that the latest scientific evidence had conclusively shown that human-made or anthropogenic climate change was real. The continued rise in global temperature, it was predicted, would cause a thermal expansion of the oceans and melt glaciers and polar ice, thus causing sea levels to rise by one to four feet by the middle of the next century.[8] A host of very negative climatological impacts were likely. The mathematical models that had predicted for some years that a buildup of carbon dioxide from the burning of fossil fuels would cause the earth's surface to warm by trapping infrared radiation from the sun, turning the entire earth into a kind of greenhouse, were proven correct according to this important and groundbreaking testimony. Senator Timothy E. Wirth, the Colorado Democrat who presided at the hearing, said, "As I read it, the scientific evidence is compelling: the global climate is changing as the earth's atmosphere gets warmer. Now, the Congress must begin to consider how we are going to slow or halt that warming trend and how we are going to cope with the changes that may already be inevitable."[9] If only it had been that simple.

Today, three decades after James Hansen and other leading scientists testified on that warm June day in 1988, the debate about climate change remains unresolved within our political and policy arenas. One finds oneself asking the same questions over and over again: If the science on global warming is settled, what is to be made of the considerable public and political disagreement in the United States over the subject? Why, if the preponderance of scientific evidence and the consensus of climate scientists is to be believed, is there such intense disagreement in political circles and among the public in general about the need to respond to the warnings that the science has so clearly demonstrated?

It is a given that, when policy issues have high stakes, political debate in the United States will be contentious. Because the potential impacts of global warming are so threatening, because the fossil fuel consumption that contributes to it is so important to the world economy, and because the costs of addressing it and/or ignoring it are so very great, strongly felt and opposing views will be expressed and argued in our public discourse. In fact, the number and the intensity of contradictory claims advanced about climate change are extreme. The average citizen neglects to consider whether these claims and counterclaims are based on the best and most reliable scientific evidence. Absent any significant background or familiarity with the scientific basis for evaluation, the average citizen is frequently left with the mistaken impression that there are two sides to the climate debate that are more or less equal and that the science itself is in dispute. But the science is not in dispute. It has steadily advanced over a considerable period, and its conclusions have become increasingly certain. Climate change is not a new invention or a liberal tree-hugger political conspiracy. It can only be understood in the context of a long history of scientific inquiry.

It was a 19th-century French mathematician, Jean-Baptiste Joseph Fourier, who first asked the most basic of all questions: Why, if it is constantly absorbing energy from the sun, does the earth not heat up until it is as warm as the sun? The answer is that the earth and its atmosphere reflect or radiate energy back out into space, thereby offsetting energy absorbed from the sun. But, of course, it does not reflect all the energy back into space. If it did, the planet would be much too cold for human life. Fourier reasoned that some heat must be retained in the atmosphere. It does not all reflect back into space because some of it is absorbed (trapped) by atmospheric gases. In other words, incoming solar radiation (emitted on short wavelengths) passes freely through the earth's atmosphere to the earth's surface, but some outgoing terrestrial radiation (emitted on a longer wavelength) is trapped by the earth's atmosphere. The gases in the earth's atmosphere act as a sort of blanket insulating the planet, causing an increase in the surface temperature of the planet. This is much the same as when you cover up with a blanket. It traps heat, and you warm up. Remove the blanket, and you cool down. It is in just this sense that some of the earth's gases act as a blanket. This is what we call the greenhouse effect.[10]

Many gases exhibit greenhouse or blanketing properties. Some of them occur in nature (water vapor, carbon dioxide, methane, and nitrous oxide), while others are exclusively human made (like gases used for aerosols). Water vapor (H_2O) is responsible for approximately two-thirds of the greenhouse effect. The largest contributor to the greenhouse effect after water vapor is carbon dioxide (CO_2) followed by methane (CH_4).[11]

The first direct measurements of the absorptive capacities of the earth's atmosphere were provided by the experimental physicist John Tyndall in 1859. Tyndall was actually able to determine the degree to which specific gases present in the earth's atmosphere are able to absorb or transmit radiation. He found that the most common gases, oxygen and nitrogen, are transparent to both solar (short-wave) and terrestrial (long-wave) radiation. But other gases, such as water vapor, CO_2, and CH_4, are transparent to radiation emitted by the sun (incoming) but opaque to radiation emitted by the earth (outgoing). In other words, they react differently or selectively, and, with respect to terrestrial radiation, some heat is trapped in the earth's atmosphere by the blanketing effect of CO_2 and other greenhouse gases. This proved that the greenhouse effect, though not yet named as such, is real.[12]

In 1893, a Swedish physicist named Svante August Arrhenius was the first to actually measure the influence on ground temperatures of carbonic acid in the air.[13] Arrhenius's exhaustive computations, confirmed by subsequent study over the next hundred years, showed that significant changes in atmospheric CO_2 levels would indeed lead to some very profound shifts in global climate. For example, he demonstrated that a reduction of CO_2 levels by half (based on the levels at his time) would lower the average global temperature by 7° to 9°C. This would be enough to cause a new global ice age. Conversely, the doubling of these CO_2 levels would cause increases in average global temperatures of between 9° to 11°C. This, he said, would cause glaciers to retreat and sea levels to rise and result in considerable global warming. Arrhenius believed that a warming trend was the most likely outcome for the future based on the pace of industrialization during his lifetime and the increasing reliance on fossil fuel combustion for energy.

The linkage between the greenhouse effect and fossil fuel consumption has been well established for over 100 years, but the question of whether human activities were enhancing the greenhouse effect in a major way would remain a subject of credible debate until the late 20th century. The ultimate answer to this question began to develop during the 1950s through the work of Charles David Keeling. Keeling, a research scientist at the Scripps Institution of Oceanography in California, installed one of the first manometers (a device designed to measure atmospheric carbon dioxide) at the Mauna Loa Observatory in Hawaii. Keeling ultimately expanded his CO_2 research to diverse areas, such as the Big Sur near Monterey, California; the Olympic Peninsula in Washington; and the mountains of Arizona. As his various monitoring stations have gathered data over the years, much has been learned that should inform our climate-change discussion today.[14]

Keeling was the first to discover the seasonal rhythm of CO_2 levels. In Mauna Loa in 1958, he observed that CO_2 levels peaked in May and then dropped to a yearly low in October. Repeated observation of this pattern in subsequent years led him to conclude that he was observing the withdrawing of carbon from the air for plant growth during the summer and returning of it each winter. This was a natural cycle.[15] Keeling was also able to determine whether the concentration of carbon dioxide in the atmosphere was uniform across the earth. The Mauna Loa manometer confirmed that it was. This meant that higher emissions in any portion of the planet would ultimately increase CO_2 concentrations in all regions of the planet. Over time, a much more profound and indisputable discovery would be made. In addition to being globally diffuse, CO_2 levels were found to be steadily and quickly rising due primarily to human influences.[16] Keeling observed that the amount of CO_2 in the earth's atmosphere was steadily increasing. This, he was able to demonstrate, was connected to the combustion of fossil fuels. The past five decades have not only confirmed Keeling's findings, but, also, climate scientists have developed an indisputable body of evidence confirming that the earth is warming and the climate is changing and that human activity is responsible for this change. Irrefutable evidence from around the world—including extreme weather events, record temperatures, retreating glaciers, and rising sea levels—all point to the fact that global warming is happening now and at rates much faster than previously thought.

Over the past 40 to 50 years, the rate at which CO_2 levels have been increasing is considered to be unprecedented in human history. In fact, the concentration of CO_2 in the atmosphere is the highest it has been in over 800,000 years.[17] According to climate researchers, the "safe" or upper level of carbon dioxide in the atmosphere should not exceed 350 parts per million. "Safe" here means conditions that will minimize the likelihood of the most severe negative effects from a significant climate change in the form of rising seas, wildfires, and extreme weather of all kinds. In early 2013, CO_2 levels reached 400 parts per million for the first time. By 2016, 400 parts per million became permanent with the level expected to continue rising significantly.

Scientists have shown the indisputable correlation between CO_2 levels and higher temperatures. In other words, the evidence is conclusive that rising CO_2 levels are correlated with global warming. The human contribution to the problem of increased CO_2 levels is the result of our dependence on fossil fuels for energy, which means the burning of oil, gas, and coal. During the past 20 years, about three-quarters of human-made carbon dioxide emissions (and these are still increasing) were from burning fossil fuels. It can be said that we, through our energy development and consumption patterns, have contributed to the creation of a dangerous imbalance in the global carbon cycle. This imbalance means that carbon is accumulating more rapidly in the atmosphere than it is being removed through absorption. This excess stock of carbon dioxide enhances the greenhouse effect and warms the climate.[18] As Keeling concluded long ago, this translates into a growing blanket of carbon dioxide, which raises the earth's average temperatures.

Whether or not one wishes to accept the scientific evidence that human activities are the main contributing factor to global warming, there is ample evidence that the earth's climate is in fact warming. Among the things that can be directly linked to increasing global temperatures are the widespread melting of glaciers and ice sheets and decreases in Northern Hemisphere snow cover and Arctic ice. The permafrost is thawing, lakes and rivers are freezing later and melting earlier, and agricultural growing seasons are being altered in some locations.

The evidence for fairly rapid climate change on a global scale is compelling. Global temperature, according to all major surface temperature reconstructions, has warmed since the 1880s. The warmest years in recorded human history have occurred since the 1970s. In fact, over the past several years, new heat records have been set in each successive year.[19] Since the beginning of the 2000s, we have witnessed a decline in solar output, yet surface temperatures have continued to increase.[20] The oceans have absorbed much of this increased heat, resulting in increasing water temperatures. The shrinking of the ice sheets has been stunning. The ice sheets in Greenland and the Antarctic have decreased in mass (36–60 cubic miles of ice per year in Greenland), and both the extent and the thickness of Arctic sea ice has declined rapidly for several decades.[21] Glaciers are in retreat, and the retreat is accelerating noticeably almost everywhere, including the Alps, Himalayas, Andes, Rockies, Alaska, and Africa.[22] Extreme weather–related events are increasing rapidly as well. The number of record-high temperatures in the United States has been increasing, while the number of record-low temperatures has been decreasing since 1950. Additionally, the increasing number and intensity of natural disasters in the United States can be linked to climate change.[23]

It is perhaps useful to consider, however briefly, the impacts of climate change on human populations *already* measured. In many parts of the world, the increasing impact of climate variability is making it clear exactly what climate change means for people and the communities in which they live. The Global Climate Risk Index, developed and compiled by Germanwatch.org, quantified in terms of both fatalities and economic losses the impacts of extreme weather events between 1996 and 2015. This database summarizes the impact of extreme weather events associated with a warming global climate and is regarded worldwide as one of the most reliable and complete databases on this matter. Between 1996 and 2015, a 19-year period, more than 528,000 people died in extreme weather related to climate variability worldwide. Economic losses of $3.08 trillion were incurred as a direct result of almost 11,000 extreme weather events.[24] Perhaps more importantly, the final report estimates escalating impacts and resulting increases in human and global costs by

2030 or 2050. The United Nations Environmental Program Adaptation Gap Report of 2016 warns of increasingly severe global impacts and costs, likely to be two to three times higher than current global estimates by 2030 and potentially four to five times higher by 2050. On the other hand, the report highlights the importance of enhanced mitigation action, such as limiting global temperature increase to below 2°C to avoid substantive costs and hardships.[25]

More general examples of climate-change impacts include increases in global surface temperature (global warming), changes in rainfall patterns, and changes in the frequency of extreme weather events. Changes in climate may be due to natural causes (e.g., changes in the sun's output) or to human activities and the changing the composition of the atmosphere. That is beyond dispute. Likewise, any human-induced changes in climate will occur against a background of natural climatic variations. But it is also indisputable that the climate is changing due to anthropogenic or human influences as opposed to natural processes. Indeed, science has been able to eliminate natural causes and to reach the conclusion or consensus that human-induced climate change is real. There is simply no doubting the impacts of climate change on humans and on natural systems generally, including on oceans, weather extremes, health, infrastructure, air quality, natural habitats, agriculture, and air quality.

Scientists are confident that global temperatures will continue to rise for decades to come, largely due to greenhouse gases produced by human activities. The United Nations Intergovernmental Panel on Climate Change, which includes more than 1,300 scientists from the United States and other countries, forecasts a temperature rise of 2.5° to 10°F over the next century. The extent of climate-change effects on individual regions will vary over time and with the ability of different societal and environmental systems to mitigate or adapt to change. Increases in global mean temperature of less than 1.8° to 5.4°F (1° to 3°C) above 1990 levels will produce a few beneficial impacts in some regions and many very harmful ones in others. Net annual costs will increase significantly over time as global temperatures increase.[26] The magnitude of climate change beyond the next few decades depends primarily on the amount of heat-trapping gases emitted globally and how sensitive Earth's climate will be to those emissions.

Scientists are able to make several general predictions as temperatures will continue to rise. There will be changes in precipitation patterns with more winter and spring precipitation for the northern United States, and less for the Southwest, over this century. Projections of future climate in the United States suggest that the recent trend toward increased heavy precipitation events will continue. There will be more severe droughts and heat waves. Droughts in the Southwest and heat waves everywhere are projected to become more intense and cold waves less intense everywhere. Hurricanes and tropical events will be stronger and more intense. The intensity, frequency, and duration of North Atlantic hurricanes, and the frequency of Category 4 and 5 hurricanes have all increased since the early 1980s. Hurricane-associated storm intensity and rainfall rates are projected to increase as the climate continues to warm. Global sea level has already risen by about eight inches since reliable record keeping began in 1880 and is projected to rise another one to four feet by 2100. Sea-level rise will not stop in 2100 because the oceans take a very long time to respond to warmer conditions at the earth's surface. Ocean waters will therefore continue to warm, and sea level will continue to rise for many centuries at rates equal to or higher than that of the current century.[27] The possibility for even more devastating impacts is greater than many might be aware, including those who recognize the reality of anthropogenic climate change.

Climate change is being experienced particularly intensely in the Arctic, although few people outside of the scientific community are paying serious attention. In fact, Arctic average temperature has risen at almost twice the rate as that of the rest of the world over the past few decades. This has contributed to the widespread melting of glaciers and sea ice that has been linked to climate change. Rising permafrost temperatures present additional evidence of strong warming in the Arctic. These changes in the Arctic provide an early indication of the environmental and societal significance of some severe global consequences. The Arctic provides important natural resources to the rest of the world (such as oil, gas, and fish) that will be affected by climate change, and the melting of Arctic glaciers is one of the factors contributing to sea-level rise around the globe. These climatic trends are projected to accelerate due to ongoing increases in concentrations of greenhouse gases in Earth's atmosphere. These Arctic changes will, in turn, impact the planet as a whole. That is not as well understood perhaps and not as broadly known as it should be.

The most threatening implication of Arctic warming has to do with the release of methane from the Arctic floor as the ice melts. Scientists have expressed some serious concerns about what could happen as methane bubbles up from the bottom because of the relatively fast melting of the Arctic ice.[28] What might this mean? As the amount of Arctic sea ice declines, which it is already doing at an unprecedented rate, the thawing of offshore permafrost releases methane gases from the ocean bottom. There is a 50-gigatonne reservoir of methane stored in the form of hydrates on the East Siberian Arctic Shelf. As the seabed warms, either steadily over 50 years or suddenly over a shorter period, this methane will be released. Higher methane concentrations in the atmosphere will accelerate global warming and hasten local changes in the Arctic. This could include speeding up the retreat of sea ice, reducing the reflection of solar energy, and accelerating the melting of the Greenland ice sheet. The ramifications will be felt globally.[29] If global average temperature were to rise, say 2.5°F (1.5°C), above where it stood in preindustrial times, and it will, permafrost across much of northern Canada and Siberia could start to weaken and decay. Scientists tell us that because of this, global warming could begin to accelerate much more rapidly than current projections due to what is referred to as a feedback mechanism. Scientists describe the feedback mechanism like this: Thawing permafrost will release larger amounts of carbon dioxide and methane, which will lead to more rapidly rising global temperatures, which will lead to further permafrost thawing, which will lead to rising global temperatures. And that trend is projected to continue through the rest of the century. There is much concern that the release rate of methane in the Arctic could accelerate. This, in turn, could lead to an acceleration of global warming with the most serious of consequences.

How much acceleration and additional damage an Arctic methane release may cause will depend on how quickly the permafrost melts and how quickly bacteria convert the plant material into carbon dioxide and methane gas. Nobody knows the full answer to that. But, as we have seen, climate scientists already expect a wide range of negative consequences from rising temperatures, including higher sea level, more weather extremes, and increasing risks to human health. Thus, anything that accelerates the rate of warming is a serious concern. Current projections indicate a 30 to 70 percent decline in near-surface permafrost by the end of this century.[30] A 30 percent decline in permafrost would present serious problems, but anything higher would be catastrophic. If 70 percent of the permafrost were to thaw, scientists expect that 130 to 160 billion tons of carbon will be released into the atmosphere by the end of this century. To put that in perspective, the United States currently emits about 1.4 billion tons of carbon annually from fossil fuel combustion and cement production.

Science is telling us a good deal. A mere listing of the already-known impacts of climate change that science has been able to tell us about is daunting: killer heat waves, torrential rains and flooding, drought, wildfires, rising sea levels, shrinking snow packs, vanishing glaciers, melting permafrost, damage to coral reef, species extinctions, new disease outbreaks, and, as a worst-case scenario, human extinction. While the last item on this list is seen as unlikely for the moment, all the others are already happening. Already impacted are things that we depend on and value: water, food supplies, energy, transportation, wildlife, agriculture, ecosystems, and human health. Scientists and economists are also beginning to grapple with the serious economic and environmental consequences that are inevitable if we fail to reduce global carbon emissions quickly and deeply. The most expensive thing we can do is nothing. The cumulative costs of damage to property and infrastructure, lost productivity, mass climate migration, security threats, and coping costs mean that trillions of dollars are on the line. Surely, given what is known and what can be reasonably projected, one would think that climate change is an important item (if not the most important) on the American policy agenda. That would be assuming, of course, that the expert knowledge and advice of climate science has any impact whatsoever. Sadly, science and politics are colliding more than they are collaborating on climate change. In fact, the collision has become more intense and dangerous even as the science has become more certain.

Climate Politics

The United States was created during the Age of Reason. As mentioned in chapter 1, it is thus no surprise that the Founding Fathers had a passion for science. Fast-forward to January 2017 and both the incoming U.S. president and vice president, along with roughly the entire cabinet and most of the majority party in Congress, are active deniers of the most well-established science of our time. The incoming leader of the free world asserted that "nobody knows for sure" if climate change is real. In light of what science has actually enabled us to know, this assertion borders on the lunatic fringe of what we might call the public discourse. Be that as it may, the lunatic fringe has competed with and often defeated science in the battle to shape public perceptions about climate change.

During his 2016 campaign for the U.S. presidency, Donald Trump referred to climate change as a hoax perpetrated by the Chinese. He would later describe this talk as a joke, but during a town hall in New Hampshire, he absolutely mocked the idea of global warming. At that event, a volunteer for the League of Conservation Voters asked Mr. Trump what he would do to address global warming. His response was pure denial: "Let me ask you this—take it easy, fellas—how many people here believe in global warming? Do you believe in global warming?" After asking three times "Who believes in global warming?" and soliciting a show of hands, the soon-to-be president of the United States concluded that "nobody" believed climate change was real. Once elected president, responding to an interview question about climate change, he said he was "open minded" about climate change. He also said, "Nobody really knows. . . . Look, I'm somebody that gets it, and nobody really knows. It's not something that's so hard and fast."[31] Many in the scientific community felt compelled to respond to such nonsensical utterings as the new president vowed to dismantle U.S. climate and clean energy policies. They weren't "nobody," and they absolutely knew. They were no doubt even more alarmed as he appointed climate deniers with ties to the fossil fuel industry to his transition team and to his cabinet. The Trump team was filled with individuals who had a proven history of attacking climate scientists and undermining climate science.

More than 800 earth science and energy experts in 46 states signed an open letter to Donald Trump, urging him to take six key steps to address climate change to help protect "America's economy, national security, and public health and safety." This was an extraordinary thing to do at the beginning of a new administration. Scientists are typically not political. They do not often write open letters expressing overt policy concerns at the beginning of a new political regime. All the signatories to this open letter either held or were pursuing a PhD in relevant scientific disciplines, and all were either American or worked within the U.S. scientific community. This open letter took a respectful tone as it offered advice to the incoming president and urged him to make the United States a clean energy user, to reduce carbon pollution and U.S. dependence on fossil fuels, to enhance the nation's climate preparedness, to acknowledge that climate change is real, to protect scientific integrity in policy making, and to uphold the United States' commitment to the Paris climate accord. (The full text of this letter may be accessed at https://blogs.scientificamerican. com/observations/an-open-letter-from-scientists-to-president-elect -trump-on-climate-change/.)

At about the same time the open letter was published in *Scientific American*, the Trump transition team sparked another fire that generated a response from the scientific community; they delivered a 74-question document to the Department of Energy, asking for the names of all personnel who had worked on climate change or attended UN climate talks within the last five years. The implied purpose of this questionnaire was, to say the least, chilling. Trump and his team vowed to dismantle specific aspects of Obama's climate policies and proceeded to do so. The questionnaire, which one Energy Department official described as unusually intrusive and a matter for departmental lawyers, raised concerns that the incoming president and his team were trying to figure out how to target the people, including career civil servants, who had helped implement policies under Obama.[32] Almost immediately after the letter's delivery, the Union of Concerned Scientists issued a press release that condemned what it said was an effort to single out specific government employees whose work included climate research. This was seen as a move to dismantle federal climate-science research altogether.

In a separate but related instance, Trump transition officials reportedly stated that they intended to eliminate NASA's climate research funding. Such a move, climate scientists quickly warned, would send us back to the dark ages. Speculation that the incoming Trump administration's Environmental Protection Agency's transition team intended to remove some climate data from the agency's Web site proved to be true. These deletions included references to President Barack Obama's June 2013 Climate Action Plan and the strategies first articulated in 2014 and 2015 to cut methane emissions.[33] Scientists, archivists, and librarians were soon hunched over laptops attempting to retrieve and preserve research that might be deleted, altered, or removed from the public domain by the incoming Trump administration. They reviewed hundreds of government web pages and data sets in an effort to preserve information strategically chosen from the pages of the Environmental Protection Agency and the National Oceanic and Atmospheric Administration.[34]

The early days of the Trump administration were marked by direct and consistent confrontations with the scientific community. No area of science has been targeted more explicitly than climate science. The new administration routinely misled the public by saying that the science around climate change isn't settled, all while implementing policies to purge the web of the actual science. The removal of data and information about climate change from federal agency Web sites deprives the public, including teachers and students, of valuable information regarding climate change. From the very beginning, the new administration implemented its strategy for derailing federal climate science and climate policy initiatives.

The moment Donald Trump took the oath as president, the new Trump White House Web site replaced the Obama site. The page devoted to climate change action was eliminated and replaced with President Trump's pledge to undo environmental regulations and "revive America's coal industry." The 361-word policy outline on the new page, titled "America First Energy Plan," made no reference to global warming or climate change at all except to note Trump's commitment to "eliminating harmful and unnecessary policies such as the Climate Action Plan and the Waters of the U.S. rule."[35] With respect to climate change, science and politics were definitely colliding as 2017 opened and a new president took the reins. By June 1, 2017, this collision resulted in the Trump administration's decision to withdraw the United States from the 2015 Paris Climate Accord. A new era of intense political conflict over climate change was at hand. It would no doubt move into uncharted and previously untraveled territory. Some feared the depth to which this conflict, if taken to extremes, would injure American democracy itself. But this conflict was not something new so much as an escalation in a long, drawn-out battle. It has been a battle that has worked against the United States' best interests for decades.

The first U.S. president to reference climate change, if only indirectly, was Lyndon B. Johnson. In 1965, the president's Science Advisory Committee produced a summary of the potential impacts of carbon dioxide on climate, which reached the conclusion that increasing levels of CO_2 in our atmosphere would lead to modification of the "heat balance of the atmosphere." This, it was reported, could contribute to significant changes in climate.[36] This report was referenced in a Special Message to Congress delivered by President Johnson in February of 1965. In reference to the scientific conclusions, he stated, "This generation has altered the composition of the atmosphere on a global scale through radioactive materials and a steady increase in carbon dioxide from the burning of fossil fuels."[37] Buried within a larger package of environmental policy initiatives, the mention of the impact of CO_2 produced by fossil fuels on the atmosphere was little noted. But it was not disputed either. Through most of the next decade, it received little attention in the policy arena.

In 1977, during the administration of Jimmy Carter, climate change emerged as a more serious concern. The Department of Energy (DOE) had reviewed its research programs related to CO_2, including the possible impacts of changing CO_2 levels on climate. Over the next couple of years, this DOE review reached the conclusion that increases in the carbon dioxide concentration of the atmosphere would result in an increase of average surface temperatures. Of greater concern than the general temperature increase associated with rising CO_2 levels was the likelihood that the warming experienced would be greater, perhaps a good deal greater, at the poles.[38] President Carter's science advisor followed up by asking the National Academy of Sciences (NAS) to convene a panel to review the findings of the DOE review. The conclusions reached by the NAS panel agreed that the climate is indeed sensitive to changing levels of CO_2 (i.e., warming is likely with rising CO_2 levels). It stated that within several decades' time, approximately 50 (perhaps fewer) years, the effects would be pronounced and have major negative impacts on the climate.[39] This study led to requests for more analysis and information, especially with respect to the timing of any potential negative impacts. A demand for more study and more scientific precision was logical and desirable.

It seemed that science and politics were collaborating very nicely in the assessment of climate change. In 1978, the U.S. Congress joined the conversation by enacting the National Climate Act, which included significant funding for climate research and authorized the National Academy of Sciences to undertake a comprehensive study of CO_2 and climate.[40] Despite this beginning, inevitable conflicts soon emerged to impede the collaboration of science and politics. Scientists were increasingly in agreement that something needed to be done about global warming, but others, for reasons that had nothing to do with science, did not agree. Short-term economic concerns and immediate material interests tend to outweigh long-term risks and vulnerabilities in our thinking. Unless a predictable hazard or disaster is at our doorstep, we defer thinking about it. It seems unnecessary in the present moment and, after all, who really knows what tomorrow will bring. This sort of thinking was the norm through most of the 1970s and 1980s as the climate change discussion heated up.

According to the NAS report funded by Congress in 1978, scientists concluded that the most likely scenario was a doubling of CO_2 in the atmosphere by 2065. They added that it would be unwise to exclude the possibility that this would happen earlier, in the first half of the 21st century. They were convinced that there would be observable and negative climate impacts.[41] It is true that the climate is naturally variable, said the scientists, but the rapid and forced change being projected due to CO_2 emissions related to increasing fossil fuel consumption are human (i.e., anthropogenic) causes that have the potential to seriously challenge ecosystems in just a few decades' time and adversely impact human life. This was the primary concern of the scientists. To address the threats that climate change would cause, things like increased taxes on fossil fuels and other regulatory measures to discourage and ultimately reduce fossil fuel consumption and carbon emissions were mentioned. The other options included adaptation to a world with higher CO_2 and its higher temperatures. Scientists preferred efforts to reduce carbon emissions.[42]

Economists who reviewed the NAS report reached very different conclusions about what should be done. They did not disagree with the scientific facts, but they did disagree with the interpretation of them. Stressing that CO_2 was not the only cause of climate change (even though the scientists were saying it was in fact *the major cause*), the economists thought it would be wrong to commit to the notion that fossil fuels and carbon dioxide were either the main cause of the problem or the area where a solution must be found. There was considerable uncertainty about both the extent and the timing of any of the negative impacts described by the science. The economists therefore thought it premature and unwise to think in terms of action, rejected as unproven the scientific conclusion that any future impacts would be potentially severe, and recommended more study and doing nothing in the near term.[43] So, as early as 1978, we had the beginning of a debate that would remain unresolved up to the present time. We have the scientists saying that climate change is a problem that poses risks to us and that we should do something to address those risks, and we have others, starting with the economists reacting to the 1978 NAS report, saying nobody really knows what will happen. The best course, these others would say over and over, is to do nothing but wait and see. The science, they would add, is far from conclusive.

Since 1978, the science has become settled. A scientific consensus has been achieved with respect to global warming, its causes, its already observed impacts, its anticipated or future impacts, and about the need to act. But, as the science has advanced and the evidence has led to scientific consensus, the climate debate has evolved into a contest between the conclusions of science and the devaluing, questioning, doubting, and ultimately denial of science for the explicit purpose of protecting economic and other interests that feel threatened by any actions to respond to the scientific warnings. Over time, the strategy of politicizing the science, denying the science, even demonizing the science in order to win the political debate has dominated the public discourse. As the science became more settled, the work to discredit it accelerated and resulted in what some would call a war against science.

Who fought this war against climate science? Well, one might begin by examining the fossil fuel industry. ExxonMobil, the world's biggest oil company, knew as early as 1981 that anthropogenic climate change was real. This was seven years before it became a public issue. Despite this, ExxonMobil spent millions of dollars over the next 27 years to promote climate change denial. Over the years, Exxon is said to have spent more than $30 million on think tanks and researchers that promote climate change denial.[44] As an example, ExxonMobil lobbied against the Kyoto Protocol in 1998 on the grounds that it would be too expensive to implement, placed too much burden on developed nations, and would have devastating economic impacts. Lee Raymond, chief executive of Exxon-Mobil at the time, was also convinced that the science behind global warming had to be wrong, or at least he perceived it to be such. Corporate discourse about climate change includes the funding of "research groups" to challenge the science.[45] In addition to funding climate denial research, ExxonMobil has donated millions of dollars in campaign contributions to politicians who deny climate change.[46]

It is not just fossil fuel companies who fund climate change denial. The largest, most-consistent money supporting the climate denial movement comes from a number of well-funded conservative foundations built with so-called "dark money," or concealed donations. These foundations, all of which promote a conservative political and corporate agenda, are not required under U.S. law to reveal the names of their contributors. This makes it possible for groups and individuals to publicly back off of direct funding of climate change denial and to funnel additional monies secretly. So it is that anonymous billionaires and front groups (i.e., 140 foundations) funneled $558 million to almost 100 climate denial organizations between 2003 and 2010.[47] The climate denial think tank Heartland Institute is one example of where the money goes. Hartland received funding from at least 19 publicly traded corporations in 2010 and 2011. The combined contributions exceeded $1.3 million for an array of projects during that one-year period, including a secret plan to teach children that climate change is a hoax.[48]

Global-warming denial or skeptic organizations, generously funded by fossil fuel, corporate, and conservative interests, are actively working to sow doubt about the facts of global warming. These organizations are key players in the fossil fuel industry's "disinformation playbook," a strategy designed to confuse the public about global warming and delay action on climate change. Why are they doing this? The answer is fairly simple: because the fossil fuel industry wants to sell more coal, oil, and gas. Even though science clearly shows that the resulting carbon emissions threaten our planet, they are pursuing their economic self-interest ahead of all else. The list of climate-change-denial groups is seemingly endless, and their agenda and tactics have become well known to most of us. One must remember that everything these groups promote is aimed at misinformation and public confusion to prevent any policy action to address climate change. Prominent climate-change denial or skeptic organizations include the American Enterprise Institute, the American Legislative Exchange Council, the CATO Institute, and the Heritage Foundation. Such organizations, funded by corporate sponsors, work in almost every way imaginable to undermine the credibility of climate science.

Why do corporations and conservative foundations seek to discredit climate science and deter any policy initiatives to address it? Corporations in the United States have always taken part in national discussions on laws and regulations that might affect their industry. In a democracy, this is their right, and nobody would dispute that. But when new scientific data reveals a threat to public health, safety, or the environment caused by their products or activities, the affected industries often use extreme and dishonest tactics to oppose any calls for regulation. They frequently do this by attacking the science upon which such regulations or policies are based. The game plan they follow is pretty much the same in each case. Corporate interests spend huge amounts of money to question the legitimate scientific consensus around an issue and to counter established findings by promoting their own studies—conducted with flawed methodologies—that lead to their self-interested and predetermined outcome. They pay seemingly independent scientists to conduct this "research" and/or to further undermine legitimate findings in media campaigns and publicity efforts. They sometimes intimidate or openly attack scientific researchers. They relentlessly skew the analysis of the costs and benefits of any proposed regulations, do everything they can to undermine the legitimacy of the regulatory process itself, and make huge campaign contributions to elect politicians who will do their bidding in the policy process. This corporate strategy has been repeated time and time again. It was first exposed in the now-infamous case of the tobacco industry's attempts to delay the regulation of cigarettes by spreading doubt about the link between smoking and lung cancer. The same tactics are being effectively used now to deter or prevent climate change policy initiatives.

Self-interested corporations and conservative ideologues who oppose all forms of governmental regulation will understandably work to discredit climate science when its conclusions support arguments for policies they find objectionable. This discrediting of science includes, as a political necessity, the cultivation of public doubt about the legitimacy of scientific conclusions, which has worked well enough to slow the policy process considerably. There is near-unanimous agreement (97 to 99 percent) among climate scientists (i.e., the active researchers and experts in the field) about anthropogenic climate change. This is to say that there is a consensus among those conducting the important peer-reviewed research in this field. Yet, for too many years, a majority of Americans have not been aware of this consensus.[49] Not only has public doubt or unawareness about climate science been cultivated but also public opinion is increasingly divided along partisan ideological lines (see Table 3.1). A Pew Research Center survey taken just before the 2016 presidential election shows the partisan distribution quite vividly. The largest differences of opinion are between conservative Republicans and liberal Democrats. In other words, political partisanship or ideology is the variable that explains most of the distribution of public opinion on this issue.

Climate change has become a wedge issue dividing the nation along partisan lines. The political fissures on climate issues extend well beyond the beliefs people have about whether climate change is occurring. The divisions reach down to the basic trust people have about researchers' motivations and the credibility of the work that climate scientists produce. What is the explanation for such stark partisan differences of

Table 3.1　Public Attitudes About Climate Scientists

Climate scientists know	Conservative Republican	Liberal Democrat	Total Pop.
Whether climate change is occurring	18%	68%	33%
Causes of climate change	11%	54%	28%
Best ways to address climate change	8%	36%	19%
Scientific consensus/ trust	**Conservative Republican**	**Liberal Democrat**	**Total Pop.**
Climate scientists agree that human behavior is responsible for climate change	13%	55%	27%
Climate scientists can be trusted to be accurate	15%	70%	39%
Climate science research findings are influenced by	**Conservative Republican**	**Liberal Democrat**	**Total Pop.**
Best scientific evidence	9%	55%	32%
Concern for public interest	7%	41%	23%
Desire to advance their careers	16%	57%	36%
Political leanings or biases	54%	16%	36%

Source: Pew Research Center, September 30, 2016, http://www.pewinternet.org/2016/10/04/the-politics-of-climate/

opinion? Is it all about political ideology? Don't educational levels, as one might normally expect, come into play?

For Democrats, ideology actually doesn't appear to be a factor because the main solutions to the problem (e.g., regulations and pollution taxes) do not conflict with their ideological beliefs. Thus, they more easily accept the science *as they become aware of what it says*. Their level of scientific knowledge, as it advances with more education, dramatically increases the likelihood that they understand that humans are causing global warming. Republicans' and conservatives' ideology would seem to prevent them

Table 3.2   Percent of Republicans and Democrats Who Agree That the Earth Is Warming Due to Human Activity (from "low" to "medium" to "high" levels of scientific knowledge)

Republican		Democrat	
Low	19%	Low	49%
Medium	25%	Medium	74%
High	23%	High	93%

Source: Skeptical Science, October 6, 2016, https://skepticalscience.com/pew-survey -republicans-rejecting-climate-reality.html

from accepting the scientific reality regardless of their level of scientific literacy. Republicans must square their ideological opposition to climate change policy solutions they find unacceptable with the consensus among experts that humans are causing global warming. This they appear to do by denying the expert consensus and distrusting the experts. So it is that recent surveys show no variation in Republican opinion across education levels, whereas among Democrats we find significant changes as education levels increase (see Table 3.2).

Conservative media continually reinforces climate-change denial. A 2013 study found that conservative media consumption (specifically *Fox News*) decreases viewer trust in scientists, which in turn decreases belief that climate change is happening. The study also examined previous research on this issue and concluded that the conservative media creates distrust in scientists through five main methods: (1) Presenting contrarian scientists as objective experts while presenting mainstream scientists as self-interested or biased. (2) Denigrating scientific institutions and peer-reviewed journals. (3) Equating peer-reviewed research with a politically liberal agenda. (4) Accusing climate scientists of manipulating data to fund research projects. (5) Characterizing climate science as a religion.[50] Is it any wonder that conservatives mistrust scientists?[51] Of course, their mistrust and suspicion is not the product of anything scientists are doing. It is the result of what conservatives fear about what the findings of science might imply in terms of public policy. Ever since climate change became an agenda item for world governments, conservative resistance has been a common feature.

In 1988, the United Nations Climate Program and the World Meteorological Association established the Intergovernmental Panel on Climate Change (IPCC). Its assigned task was to assess available scientific data and the possible impacts of climate change to determine if a global response is necessary to address any major and identifiable risks and vulnerabilities.[52]

As the IPCC began its work in the late 1980s and early 1990s, governments around the world also placed climate change on their agendas.
High-profile international conferences called for the reduction of worldwide carbon emissions. The goal was set at a 10 to 20 percent reduction in
the rate of carbon emissions. A Framework Convention on Climate Change
(FCCC) was signed in 1992 (the first international treaty on climate) and
went into effect in 1994. One hundred and ninety nations, including the
United States, ratified this agreement. With the goal of stabilization and
eventual reduction of greenhouse gas concentrations, the FCCC provided
a starting point for negotiations to identify and achieve more specific and
binding measures.[53] This effort culminated in the Kyoto Protocol of 1997.

The Kyoto Protocol sought to reduce emissions of greenhouse gases to
15 percent below 1990 rates by 2010. Thirty-seven industrialized nations,
including the United States, were required to reduce greenhouse emissions.[54] The Kyoto agreement essentially established emission limits but
allowed flexibility in how nations might choose to achieve these limits.
Most of the details with respect to the implementation of the protocol
were left to be settled by subsequent negotiations. Most of its goals were
never achieved, although ongoing international climate negotiations continued. In reaction to the Kyoto Protocol, U.S. global-warming skeptics
(fossil fuel interests, corporate interests, conservatives, etc.) developed a
plan of action. Oil and gas giants like ExxonMobil lobbied extensively
against the Kyoto Protocol on the grounds that it would be too expensive
to implement, placed too much burden on developed nations, and would
have devastating economic impacts. They mounted a relentless campaign
to oppose the Kyoto Protocol. The United States, in fact, never ratified the
Kyoto agreement. President George W. Bush, claiming that "science was
not certain," announced in early 2001 that he would not send the Kyoto
Protocol to Congress for ratification.[55]

With the election of Barack Obama to the presidency, there seemed to be
some hope for a more proactive policy approach to the climate challenge.
Indeed, in his first State of the Union address in 2009, President Obama
spoke of the need to address the "ravages of climate change." He called for a
market-based cap on carbon emissions. This would have capped the overall
level of carbon emissions that could be produced. Companies that exceeded
their specific emissions cap could, under this approach, lease additional
emission credits from companies that produced less than their allotted
amount. This was a market-based approach to create financial incentives to
reduce carbon emissions. Predictably, the proposal went nowhere in a
deeply divided Congress.[56] Indeed, throughout his first term in office, President Obama met uncompromising resistance from Republican members of

Congress and conservative forces. The issue of climate change had become a wedge issue in U.S. politics and would not be much mentioned or addressed by President Obama until well into his second term.

In December 2011, a Pew Research Center poll confirmed that the public remained stubbornly divided over the issue of climate change. They also remained confused about what exactly science was saying about climate change. Among those polled, 38 percent believed that human activity is the main cause of global warming. This was a modest increase over the 34 percent figure for 2010. But far more interesting is the partisan divide in public opinion. By 2011, it had become increasingly stark. Over half (51 percent) of Democrats and 40 percent of independents agreed with the scientific consensus that climate change is due primarily to human activity. Only 19 percent of Republicans and 11 percent of Tea Party Republicans agreed with the scientific consensus.[57] This disconnect between scientific knowledge and public and partisan opinion became so complete that the mere discussion of climate change in public and political discourse was deemed a liability in attracting support and winning elections. The consensus of the scientific community, the mounting evidence suggesting that climate change is the most important challenge facing humanity in the 21st century, and the objective facts of the matter had little influence on the political agenda or the conversation of the 2012 and 2016 presidential elections. In the four major presidential debates in 2012 between President Barack Obama and Republican nominee Mitt Romney, the issue of climate change never came up. Not a single question about it was posed by debate moderators, and neither candidate brought it up. In 2016, four years later, in the four major presidential debates between Democratic nominee Hillary Clinton and Republican nominee Donald Trump, the issue of climate change never came up. Not a single question about it was posed by debate moderators.

It is not that our government or the governments of the world were lacking in meaningful policy options to explore and debate. It has long been very generally agreed that the reduction of carbon and other greenhouse-gas emissions is a mitigating action that must be taken in response to climate change. The scientists have been saying this for almost 50 years, and increasingly we have seen that governments around the world are in agreement.

There have been three basic policy options most frequently discussed in the United States with respect to climate change mitigation (reducing carbon emissions): (1) a carbon tax that requires producers and consumers to bear some of the costs, (2) a cap-and-trade system that restricts the amount of carbon that can be produced and that allocates emissions through a market-based trading system, and (3) stricter and direct regulation of carbon emissions. The first two options, a carbon tax and a cap-and-trade system, seem to be the preferred options based on the assumption that the most efficient approach for emissions reductions is to give businesses and households an economic incentive.

It has become an accepted orthodoxy in the United States that the most efficient approaches to reducing emissions involve giving businesses and individuals an incentive to curb activities that produce CO_2 emissions rather than adopting a command-and-control or regulatory approach in which the government would mandate how much individual entities could emit or what technologies they should use. A 2008 Congressional Budget Office study concluded that a tax on carbon emissions would be the most efficient incentive-based option for reducing emissions.[58] A carbon tax is, in essence, a pollution tax. It imposes a fee on the production, distribution, and use of fossil fuels based on how much carbon their combustion emits. The government sets the rate per ton for carbon emissions and then translates this into a tax on the production and consumption of things like electricity, natural gas, and oil. The basic assumption behind this tax is that it will make using dirty fuels more expensive. This in turn will encourage utilities, businesses, and individuals to reduce consumption and increase energy efficiency. A carbon tax is also thought to be a means to make alternative energy more cost competitive with cheaper and polluting fuels like coal, natural gas, and oil. According to the Congressional Budget Office, a very modest carbon tax of $25 per metric ton of carbon would, over a 10-year period, reduce emissions by 10 percent and generate a trillion dollars in new federal revenues.[59]

In a cap-and-trade system, the government would set laws for maximum allowable emissions. This is the cap, and it would apply to all polluting industries. For every ton of CO_2, a polluter, a power plant for example, reduces under the cap that has been set for it, it is awarded one allowance. As an entity accumulates allowances, they can be sold, traded, or banked for the future. Any polluter that has successfully reduced emissions below its mandated level can auction off or sell its allowances to those who are overpolluting. This is a built-in cash incentive to reduce emissions that encourages compliance, promotes innovation, and enables the market to efficiently reduce emissions. A cap-and-trade system is popular with those who believe that a carbon tax is punitive and/or with those who believe, just because it is a tax, it is impossible to enact. Despite its relative popularity as an option (several cap-and-trade proposals have

been offered over the years), congressional support has never been sufficient to enact such a policy.

The third policy option, the direct regulation of carbon emissions, has
been implemented with limited success. Regulatory action by the Environmental Protection Agency has helped to reduce emissions rates. More
stringent vehicle fuel economy standards, new permitting requirements
for stationary facilities, and increased performance standards are making
a contribution, but this progress is offset by the reality that the regulatory
steps taken are relative baby steps when it comes to actually addressing
the greenhouse gas emissions problem. In addition to the partisan division over climate change, one must also take into account the increasing
public hostility to regulation in general and environmental regulation in
particular. The conventional wisdom here is that a regulatory approach
must of necessity be moderate to be acceptable or politically feasible.
Given climate change denial and conservative reactionary resistance to all
forms of regulation, there are severe limits to what regulatory bodies may
be able to achieve.

In the summer of 2013, President Obama unveiled his U.S. Climate
Action Plan,[60] consisting of a wide variety of executive actions designed to
achieve three major goals: (1) Cut carbon pollution in the United States.
But instead of legislative proposals (e.g., a carbon tax, a cap-and-trade
regime, new regulations), the plan said that the president would act
through executive orders and the existing regulatory authority of federal
agencies (e.g., Environmental Protection Agency) to put tough new rules
in place to cut carbon pollution. (2) Prepare the United States for the
impacts of climate change. With respect to the anticipated impacts of climate change, the president articulated the priority of helping state and
local governments strengthen their roads, bridges, and shorelines so they
can better protect people's homes, businesses, and way of life from severe
weather. (3) Lead international efforts to combat climate change and prepare for its impacts.[61]

President Obama knew full well that resistance from a Republican-
controlled Congress would stymie any policy initiatives, so he would
have to use the regulatory authority of the executive branch to accomplish anything. Congress has never produced anything close to a comprehensive climate and energy bill. It has never passed a significant bill to
reduce carbon emissions. Under Republican leadership, it was not likely
that it would ever do any of these things. With the inauguration of Donald Trump in January 2017, the U.S. Climate Action Plan of 2013 was also
dead. Trump's "America First Energy Plan" included getting rid of the
regulations that Obama put in place to protect the environment and

reduce carbon dioxide emissions in the United States. The new president and the Republican Congress were committed to eliminating policies and regulations aimed at protecting the environment and addressing climate change. This they did with devastating swiftness.

The World Climate Summit of 2015 may have represented the most hopeful moment in over two decades of attempts to achieve an international agreement to reduce carbon emissions. President Obama announced the U.S. plan to reduce CO_2 by 26 to 28 percent over the next decade through existing regulatory authority. In essence, this was a continuation of the Obama strategy to do what could be done through executive action. There were no new policy initiatives on the horizon and absolutely no reason to believe there would be any time soon. Kentucky senator and Republican majority leader Mitch McConnell warned the UN and the international community to "proceed with caution" in negotiating a climate deal with the United States because the country (if he and the Republican Party had their way) would not make good on Obama's commitment to reduce emissions.[62] McConnell's statement came just hours after the U.S. government submitted its pledge to the United Nations. This shows just how partisan and contentious the issue of climate change had become in the United States. Yet, even in the midst of a political effort to scuttle climate policy initiatives and to deny the basic facts of climate science, significant progress still seemed possible at the end of 2015. Even as the United States withdrew from the 2015 climate agreement in 2017, the rest of the world remained determined to pursue the goals set forth in Paris and to promote future progress.

The historic climate agreement in Paris in December 2015 was indeed a significant accomplishment, not for what it would actually accomplish, for it would be incomplete or inadequate to meet all of its objectives, but as a meaningful beginning with future progress to be made in subsequent talks and an understanding that it represented the first agreement that actually made a dent in the challenge of reducing carbon emissions. The world celebrated. President Obama, now regarding climate change as a central element of his legacy, spoke of the deal in a televised address from the White House. "This agreement sends a powerful signal that the world is fully committed to a low-carbon future," he said. "We've shown that the world has both the will and the ability to take on this challenge."[63] Scientists and leaders from around the globe agreed that the Paris climate agreement represented the world's last best hope to begin a process that would avert the most devastating effects of a warming planet.

With the end of the Obama administration and the beginning of a Trump administration, and with Republicans holding majorities in both the House and the Senate, the collision between science and politics over climate change was set to become ever more dramatic and destructive. Republicans immediately began rolling back environmental regulations and doing everything they could to maximize profits for the fossil fuel companies that had supported them with campaign contributions. With the cabinet and subcabinet appointments of the new Trump administration, fossil fuel and corporate interests now also controlled the executive branch of government, perhaps more completely than ever before in the history of the nation. Republicans in Congress meanwhile recommitted themselves to a partisan witch hunt against government climate scientists. They began with a trumped-up attack against the National Oceanic and Atmospheric Administration (NOAA). Falsely accusing NOAA of falsifying data to justify a partisan agenda, the House Science Committee pledged itself to an effort to stop scientists from deceiving the American people. Meanwhile, climate scientists are the target of a new onslaught of intimidation from inside and outside government.[64] The United States now has a president and a congressional majority fully engaged in the war against climate science.

The partisan divide over climate change had even before 2017 become so intense that climate scientists faced threats of violence, even death threats. Michael E. Mann, renowned climate scientist at Penn State, publicly shared his experiences, saying, "I've faced hostile investigations by politicians, demands for me to be fired from my job, threats against my life and even threats against my family." Mann added that "with the coming Trump administration, my colleagues and I are steeling ourselves for a renewed onslaught of intimidation."[65] One cannot help but be very concerned with what this says about the relationship between science and politics in the United States.

Conclusion

As one reviews the path from June 1988 and the testimony of James E. Hansen before the U.S. Senate Energy and Natural Resources Committee to the present day, there is little wonder that we have ended up where we are and that public surveys and polls reveal an ongoing skepticism and criticism of climate scientists as partisan alarmists. There is little doubt about why we see a general distrust of the most reliable climate data science has gathered and shared to date. It has been said in many different ways, and social scientific analysis has confirmed, that there is a natural disconnect between the reality of our perceptions and our perceptions of reality. This natural disconnect produces a mental inertia that slows down our capacity to respond efficiently to scientific knowledge. Even unbiased minds can be slow or inefficient in processing and responding to new information that challenges our values and our previously held views of reality. But when the political contest involves a concerted, well-funded, sophisticated, and ideological campaign that attacks and denies the reality that science reveals, the resulting inertia might not merely slow down our capacity to respond; it might cripple our ability to respond at all and accelerate the negative consequences of our inaction.

As social scientific analysis has shown, political partisanship or ideology is far more influential than the science with respect to the formation of public opinion about climate change. This is to say, with respect to what the public thinks about climate change, emotions, values, beliefs, and a variety of distortions or misperceptions all matter more in shaping the public discourse than the work of climate scientists. The division of public opinion and the debate about climate change in the United States is not a product of science. The science is robust, and its conclusions support a solid consensus of expert opinion about anthropogenic climate change and the risks it poses. But the debate, such as it is, is entirely political. We have seen that Democrats tend to agree with the climate scientists and Republicans tend to disagree. Why do we continue to debate climate change? It can be said that debate is necessary in science. The current scientific consensus on climate change is that global warming is a stable long-term trend, the trend is largely human-caused, serious damage is already observable, and more serious damage will result at some future date if immediate steps are not taken to halt the trend. However, there is also a very small but vocal number of scientists in climate and climate-related fields that disagree with the consensus view. In the context of scientific research and dialogue, that debate must be conducted, and it must be decided by the scientific data. The existing scientific consensus, in the context of the ongoing scientific "debate," remains undisturbed. The evidence supports it.

The fact is that most climate scientists, 97 to 99 percent of them, along with the U.S. National Academy of Sciences and more than 30 professional scientific research societies agree that climate change is happening because of human actions. They also agree that we are in serious trouble if we don't do something about it. So why indeed are we so divided? Why do so many Americans deny the reality of climate change? Some say it is because of anti-intellectualism, an attitude that minimizes the value of intelligence, knowledge, and curiosity. Anti-intellectualism is the belief that science, expertise, and "book knowledge" are less valuable than "street smarts" and common sense. Those who share this orientation also believe that they don't have to read anything about a field of knowledge before dismissing it with their own theories. There is no doubt that anti-intellectualism is a common feature in American culture, but it is a bit overly simplistic to say that it explains why the United States is so divided over climate change. The broader answer may be found in the unhealthy conflict dynamic that is at work in the relationship between science and politics.

The conflict dynamic, as we have said, is frequently the defining characteristic of issues where public policy and science must intersect. Let us recall what we have previously said. The conflict dynamic is one in which we have an issue or concern where the conclusions or recommendations of science generate strong opposition from special interests and the political entities that they fund and support. Such opposition is based most frequently on economic or material interests. This opposition becomes a part of the policy debate, often to the extent of muting or ignoring the actual science. It is natural to assume that if people do not accept the science of climate change it might be because they do not understand it, or they don't know about it. Certainly it is true that someone who knows very little about climate change is not likely to care a great deal about its consequences. The facts about climate change are widely known and readily available, yet many people say they are unconvinced that climate change is actually happening. They express more uncertainty than climate scientists do about the seriousness of the problem. This, one might suggest, is due to the concerted efforts of political, ideological, and corporate campaigns to create and reinforce doubt and denial. The creators of doubt have waged a sophisticated and very successful campaign against science.

We have seen that Republicans tend to deny the scientific consensus regarding climate change. This is no doubt because their conservative views color their interpretation of the science. They see its conclusions as threatening to their ideology. Fossil fuel companies, protecting their economic interests and conservative foundations promoting their ideological policy preferences have campaigned vigorously to animate this reaction among conservative voters and to confuse the rest of the public with distortions and distractions that question or deny the conclusions of science. Those campaigning to discredit climate science have used a variety of tactics. In addition to misrepresenting scientific data, cherry picking data, skewing the interpretation of data, or producing pseudoscience doctored to make their case, they exploit a number of other rhetorical tricks to direct the discussion away from what the science actually says. These tricks use psychological, economic, political, epistemological, and metaphysical argumentation cleverly manipulated to their advantage.

Psychologically, the consequences of climate change may seem too awful to contemplate. There may thus be a predisposition to deny the issue. We often resort to denial as a defense against unpleasant truths. If you don't look at it, it can't look at you—this is often the human reaction to things that are unpleasant or disturbing to us. Playing to this predisposition may be a small part of the overall strategy, but it is exploited, nonetheless. Misdirecting our focus with a bogus economic argument is an essential and frequently used strategy. For example, the costs of a large-scale effort to fight global warming are said to be too steep to bear. Therefore, we decide we must defer action, or we conclude that the problem doesn't exist or isn't a major concern. We also almost always accept the notion that the economy (including development) is more important than any environmental concern we may have. Politically, discrediting climate science is another common tactic. It emphasizes the fact that Democrats are always going on about climate change whereas Republicans are not. This is said to suggest that climate change is a political issue, not a scientific one. This creates a feedback loop: If climate change is real, why is it so polarizing? Because it's so polarizing, climate science is political, and we should be very suspicious of it. Epistemologically, the denier strategy simply dismisses all evidence. The question is, Why should we believe in climate change? Where's the evidence? All we know is what scientists say, and scientists are sometimes wrong or biased. And, of course, the lazy person's metaphysical argument says that God simply isn't going to let millions of people die in an epic drought.

It is safe to say that the political conflict manufactured over climate change has absolutely nothing to do with climate science. It is about discrediting climate science or creating denial of it in order to protect or promote the interests of the fossil fuel industry and to limit the capacity of the government to act. The politics of climate change have been driven by effective campaigns over many decades to mislead the public and discredit scientific knowledge. The political debate, animated by the ideological effort to deny the science, has been a false debate where the science is concerned. On one side of this "debate" are the well-established findings of legitimate science and on the other is a well-funded "hoax" of climate change denial. These two sides do not live in the same world, and they do not share the same reality. This is a particularly dangerous situation, and it does nothing but harm the broader public interest.

Everything we know about climate change, and we obviously don't know everything, tells us that there will be risks and vulnerabilities aplenty. Each region of the United States, for example, will sustain economic, environmental, and health impacts that will range from costly to devastating. Refusing to recognize and respond to these risks and vulnerabilities is, according to the science, a grave threat to the communities in which we live and work and to the human population. But the science, and we have only covered in this discussion the broad surface of what the science tells us, has neither informed nor influenced our public policy actions nearly enough. The conflict dynamic has worked against that. If anything, it has muted the science and amplified the politics of denial. Yes, small bits and pieces of progress may have been made over the past 30 years, but these baby steps have been offset and ultimately overtaken by the self-interested politics of denial. The contentious relationship between science and politics has become in our present time a persistent and effectively executed war against climate science.

Thinking scientifically, one might say the best policy options that would achieve the optimal result with respect to climate change mitigation are those that will leave fossil fuels in the ground, promote much more rapid development of alternative and cleaner energy sources, and establish the bold goal of decarbonizing our energy economy within the next 20 years. One could even argue that the economic leaders of tomorrow will be the nations that take the lead in doing these things. Politically, the United States after three decades of serious debate has retreated to policy options that the scientists believe will end in disaster. An emphasis on more fossil fuel extraction, fewer (if any) environmental regulations, and the elimination of climate change regulations and initiatives is where we have ended up.

To be fair, corporate and energy interests really do not oppose the science as much as they simply want to protect their undisturbed right to chase trillions of dollars of fossil fuel profits still in the ground. The threat this poses to people and to the earth's ecosystems are market externalities they want to simply ignore. That's business as usual. The point is, absent any policy interventions that interrupt the normal flow of their activity, energy interests will pursue profits without giving a second thought to the costly carbon externalities they produce on their way to the bank. At the present time, the prospects for any truly meaningful policy interventions that can bend the arc toward a responsible climate change mitigation effort are practically nonexistent. Since 1988, in three decades' time, we have moved scarcely an inch toward the goal of addressing climate change. We know that if we continue to use fossil fuels as our primary energy source, the conditions of life on the earth for our own species and for others will be damaged perhaps beyond repair. And yet the United States proceeds with its eyes wide shut. Our politics has crippled our ability to act, and our science has not influenced our politics nearly enough. As a result, we are doing very little to avert an impending and entirely foreseeable catastrophe.

Climate change is a long-term and complex problem. Given this, the conflict dynamic is the expected dynamic at play in the relationship between science and politics. We are generally not very future oriented in our political thinking, especially when long-term ecological concerns conflict with short-term or immediate economic and political concerns. The immediate, the short-term focus, usually wins out. But even where the scientific concerns are related to immediate phenomena or threats, the conflict dynamic animated by economic self-interest still frequently impedes the ability of science to be heard. The scientist and the politician still live in different worlds. They still do not share the same reality. The public interest still remains ill served. In the context of more immediate concerns, the collision between science and politics plays just as big a role as it does in the longer term. We turn next to another case where the conflict dynamic works to impede progress in the promotion of sound public policy.

CHAPTER FOUR

Hydraulic Fracturing: A Deepening Fissure

In March 2008, Craig and Julie Sautner had just moved to Dimock, Pennsylvania, a small community (population 1,400) very much to their liking, and they had found their dream home. As they were busy remodeling their new home, representatives from Houston-based Cabot Oil & Gas (a midsize player in the energy-exploration industry) came knocking on their door. Cabot, interested in mining the Pennsylvania shale deposits for natural gas, had come to inquire about leasing the mineral rights to their three and a half acres of land. The Sautners were told that their neighbors had already signed leases and that the drilling would have no impact whatsoever on their land.[1] Of course, under Pennsylvania law, a horizontal fracturing well drilled on a leased piece of property can capture gas from any neighboring properties. Technically, Cabot did not need the Sautners to sign a lease to capture the gas under their land. The Sautners, however, agreed to sign the lease. They sold their mineral rights for $2,500 per acre plus some small royalties on each producing well. This was actually a better deal than their neighbor across the street had received, and it was better than receiving nothing, as Cabot could have accessed the gas under their land from neighboring properties. All things considered, it seemed like the Sautners had gotten a good deal from an apparently honest company. Soon after the Sautners signed the agreement, the ground was cleared to make room for a four-acre drilling site. Operations began less than 1,000 feet away from their land. As the drilling commenced, the Sautners soon began to feel some unexpected and very unwelcomed impacts.[2]

Within the first month of drilling, the Sautners began to experience some disturbing things. Their water turned brown and became so corrosive that it scarred dishes in their dishwasher and stained their laundry. After they complained to Cabot, the company installed a water-filtration system in the basement of their home. This seemed to solve the problem, and things appeared to improve. But the Pennsylvania Department of Environmental Protection (DEP) came to do further tests in response to other complaints from local residents having similar problems. The DEP found that the Sautners' water, and the water of other local residents, contained dangerously high levels of methane. Cabot, in response to this finding, added more ad hoc pumps and filtration systems. The Sautners were advised not to drink the water at this point, but they were told that they could continue to use it for other purposes. This they did for a full year but not without new complications. Their daughter was often overcome by the dizzying effect the chemicals in the water had on her when she took her morning shower. Many mornings, as a result of this dizzying effect, she would find it necessary to get out of the shower and lie on the floor. The Pennsylvania DEP soon discovered and acknowledged that "a major contamination of the local aquifer had occurred."[3] Methane and dangerously high levels of iron and aluminum were found in the Dimock water supply. As a result, the DEP took all the water wells in the Sautners' neighborhood offline.

After the Sautners' well was taken offline, Cabot began delivering water to the Sautners' home every week. But this was of little comfort to them. The situation had become unbearable, and they soon decided they needed to move. The family could no longer take showers at home. Their concerns about the health risks associated with the chemicals contaminating the neighborhood were mounting. They were soon desperate to move, but the prospects of finding a buyer for their home were not good. Their property had become worthless. It was impossible to sell, and they could not afford to buy a new house on top of their current mortgage.[4]

The Sautners' story was not unique. Their problems were shared by many others, as private and public land in and surrounding Dimock in the watershed had been leased to energy companies eager to drill for natural gas. This drilling implemented a new technique for natural-gas extraction called horizontal hydraulic fracturing. "Fracking," as it's colloquially known, is a process that is somewhat technical and complex, but its basics are easily understood. Hydraulic fracturing involves injecting millions of gallons of water, sand, and chemicals, many of them toxic, into the earth at high pressures in order to break up or fracture rock formations and release the natural gas trapped inside. This procedure left

Dimock and the farmlands surrounding it scarred with square-shaped clearings for drilling sites and newly constructed roads for access. But this was a minor and unimportant concern as the story of Dimock would emerge in the nation's press. The chemical contamination of the aquifer that residents relied on for their fresh water would be the news for which Dimock became most known. Brown water, sick animals and people, and spontaneously combusting water wells soon made national headlines.[5]

As stories like Dimock's began to circulate, the energy industry embarked on a campaign to persuade Americans that horizontal drilling for massive new supplies of natural gas was perfectly safe. The horror stories associated with fracking being told across the nation were said to be exaggerations. ExxonMobil CEO (and future U.S. secretary of state), Rex Tillerson, led the aggressive industry-funded defense of horizontal hydraulic fracking.[6] Industry claims that hydraulic fracturing is perfectly safe, notwithstanding, lawsuits pitting residents and environmental groups against oil and gas companies have multiplied over the years, and many cite scientific studies or journals. In Oklahoma, for example, plaintiffs argued that the companies caused significant earthquakes across the state by injecting fracking wastewater into disposal wells. Geological survey presentations were introduced as evidence to substantiate these claims.[7]

Is fracking an economic boon or an environmental danger? The answer to this question is undoubtedly a bit of both. The industry says the fracking of shale gas will contribute significantly to meeting future energy needs. Is this correct? The likely answer is a qualified yes. Environmentalists say potentially carcinogenic chemicals used in the drilling process may escape and contaminate groundwater around the fracking site. Is this correct? Again, the answer is probably yes. The industry suggests that pollution incidents are the result of bad practice rather than an inherently risky technique. Is this true? It absolutely is a part of the truth. As these questions and the previous anecdotal snapshots suggest, hydraulic fracturing for natural gas and the public-policy decisions relevant to it would seem to indicate that science and politics both have a role to play in these decisions. Ideally, and avoiding for now the question of whether fracking should be a priority for our future energy policy, science provides the needed expertise for the political policy maker to adequately assess and manage or eliminate risks. But not even that much appears to be possible as our policy makers and the industry race ahead with a fracking revolution.

Shale gas is seen by many as a cheap, clean, and plentiful source of energy, a low-carbon "game changer" that will help us meet the world's rapidly growing demands for energy and offer the United States greater energy security. But the rapid rise of this new horizontal drilling technology has not been without controversy. Its popularity and profitability notwithstanding, there are serious problems to be managed. Earth tremors, surface and groundwater contamination, and the effects of fracking on human and animal health are all high-profile concerns that science has barely begun to study and address. Nevertheless, the drilling has gone full speed ahead. Drilling and fracking can be done safely, it is said, but too often it apparently is not. Careless companies spill or dump some of the 9 billion liters of contaminated water that flows back up fracked wells each day. This leads to the contamination of local waterways. Encasing wells in steel and concrete is thought to be sufficient to prevent leaks. But either as the concrete cracks over time or due to poor construction in the first place, this may inevitably allow fracking fluids to seep into drinking-water supplies or natural gas to escape. As we have seen, sending the undrinkable and contaminated wastewater back down specially permitted disposal wells has been linked to earthquakes from Ohio to Oklahoma. And methane gas can slowly and steadily leak from wellheads and pipelines, trashing the atmosphere and posing great risks to human health.

The general tenor of the debate about fracking could induce schizophrenia. "Fracking is safe," and "fracking is rampant poisoning." Both of these are, one could argue, partially true statements. In the midst of this debate, science is having a difficult time getting through. The fossil fuel industry and politicians are running dangerously far ahead of the science, discrediting it or tabling it where they feel the need to do so in the interest of profit or ideological pursuits. In the case of hydraulic fracturing for natural gas, the conflict dynamic is reinforcing the short-term focus of political and economic actors on material interests and profit. As we shall see, this works to the detriment of the public interest on a number of levels. Some would say that any hopes of imposing rigorous and needed safety and environmental controls on the industry are frustrated by the industry's political and financial influence. This influence, it might be said, brings politics and science into collision mode. To see how this might be the case and how it might harm the public interest, let us examine a little closer what might be called the fracking revolution.

The Fracking Revolution

Hydraulic fracturing is a process used in 9 out of 10 natural gas wells, of which there are over a million, in the United States. Hydraulic fracturing or "fracking" is a process where millions of gallons of water, sand, and chemicals are pumped underground to break apart the rock and release the gas. Some scientists are worried that the chemicals used in fracturing may pose a threat either underground or when waste fluids are handled and sometimes spilled on the surface. This technique for natural-gas extraction has been in use for over 50 years. Initially this was done through vertical drilling, but in the early 21st century, new technological innovations made horizontal drilling techniques feasible, which greatly expanded the potential for natural-gas extraction as well as its profitability. This new drilling technique has also introduced (predictably, of course) new risk and hazard potentials to be considered and managed.

Hydraulic fracturing has generated both staunch support and significant criticism. Supporters have applauded the new technologies that have made the accelerated exploration for natural gas possible. They have praised natural gas as a cheap, clean, abundant fuel for the future. Because it is "cleaner," it has also been hailed as a "bridge fuel" as we transition to cleaner and more renewable energy sources in our efforts to address climate change. Some critics have suggested that the new technologies associated with the horizontal drilling for natural gas will have large and undesirable environmental effects and pose significant risks to public health.[8] Indeed, concerns about the possible risks associated with fracking have escalated, as this method of natural-gas extraction has become more commonplace and its effects more widely debated.

In July 2009, the U.S. Department of Energy announced that estimated U.S. natural-gas reserves were 35 percent larger than previously estimated.[9] Much of this abundance was attributed to the new possibilities for tapping into unconventional sources, including ocean deposits of methane hydrates, coal-bed methane, and shale-gas deposits. Together, all these unconventional sources, it was thought, could provide as much as 60 percent of all natural gas in the United States by 2035. This inevitably meant that these unconventional sources would be the prime targets for accelerated application of new technologies. The most inviting target, and the one that was already being most avidly pursued, appeared to be shale-gas deposits.[10] New horizontal drilling technology made it an inviting and logical priority for energy producers.

Shale is one of the most common kinds of rock in the United States and is found in 23 states. Shale-gas deposits are plentiful, and it is estimated they may provide about 45 percent of the natural gas in the United States by 2035.[11] Accentuated by the unquestioned popularity of natural gas as a cleaner-burning fuel and as an affordable energy alternative, the large deposits of shale gasses in the United States have been rapidly developed. It is not an exaggeration to say that what has resulted might be called a revolution in natural-gas exploration.[12]

During the first decade of the 21st century, shale-gas production exploded. There was a rapid and relatively unregulated expansion of shale-gas production in the United States. It began in Texas (the Barnett Shale Field) in 2000 and has led to a race to leverage immense shale deposits around the country. The two best-known deposits may be Hainesville Shale in Louisiana and Marcellus Shale that stretches from West Virginia through Pennsylvania and New York. U.S. shale-gas production jumped from almost zero to about 2 trillion cubic feet between 2000 and 2008.[13] As shale-gas exploration took off, the conversation about the new horizontal fracking technology and any potential risks it may pose to public health and safety lagged behind the advancement of the technologies that contributed to it. Natural-gas producers, public policy makers, and media were relatively silent on these risks as production accelerated. As the boom continued, it enjoyed unquestioned support from policy makers and the public. Hydraulic fracturing was largely accepted as a good thing, its hazard potentials and risks never fully explored, and it has never been subjected to significant safety regulations.

As we have noted, the last two decades have seen advancements in technology that have taken drilling and fracturing for natural gas to new levels.[14] The major technological advancement has been related to new horizontal drilling techniques that have enabled producers to extract gas from deposits that used to be inaccessible. As drilling advanced to where drillers were able to frack horizontally, it broadened greatly the potential for extraction from a single well and improved its profitability in no small measure.[15] While popular and profitable, it is important to know that this new technique of natural-gas exploration is not without some potential and significant risks to human and animal populations. Let us take a look at an overview of the horizontal fracking process and, as we do so, examine a few of the potential risks that must be considered and managed more carefully.

The horizontal fracking process starts with drilling a hole vertically or at a slight angle from the surface to a depth of one to two miles and sometimes more. This vertical well hole is then encased in steel and/or cement to ensure that it doesn't leak into any groundwater. When the vertical well reaches the deep layer of rock where natural gas or oil exists, it is curved about 90 degrees. Drilling then continues horizontally along that rock layer, often extending more than a mile from the vertical well bore. Once a fracking well is fully drilled and encased, fracking fluids are pumped down into the well at extremely high pressure, exceeding 9,000 pounds per square inch, powerful enough to fracture the surrounding rock and create the fissures and cracks through which oil and gas can flow. The fluid that is pumped into the well to fracture the rock is called slickwater. Slickwater is mostly water but also contains a wide range of additives and chemicals, including things such as detergents, salts, acids, alcohols, lubricants, and disinfectants. These chemical cocktails usually make up no more than 2 percent (often less) of the slickwater. In addition to the water and chemical additives, sand and ceramic particles are also pumped into the fracking well. These act as "proppants" and are added to prop open the fractures that form under pressure to ensure that gas and oil will continue to flow freely out of the rock fractures after pumping pressure is released.

Once the underground rock is fractured and the proppants are pumped into place, the extraction process begins. Trapped reservoirs of gas and oil are released and pumped back to the surface along with millions of gallons of the fracking fluids, called flowback liquid. The retrieved gas is piped to compressor stations, purified, and compressed for transport. The returning fracking fluids or flowback, now called wastewater, are handled in a variety of ways. They may be transported to water treatment plants (which are not really designed to handle or treat fracking fluids), they may be stored in large tarp-lined pits and be allowed to evaporate, or they may be reinjected into old wells.[16] As one might expect, fracking fluids are a primary source of concern because of their chemical composition, usage and disposal, possibility of chemical or waste spills during transportation, potential risk of polluting water tables needed for drinking water and agricultural use, and other potential public health-related impacts.

Many fracking fluids are toxic to humans and wildlife and include chemicals known to cause cancer. Chemicals used in fracking—and there are variations as each company brews its own chemical cocktail—may include benzene, toluene, boric acid, xylene, diesel fuel, methanol, formaldehyde, and ammonium bisulfate.[17] Drillers are not required to report or make public the formulas for the chemical cocktails they create and use in order to protect their proprietary interests. The potential for the contamination of groundwater from these chemicals exists primarily from the possibility of leaks through cement well casings (over time, these will deteriorate and crack). Most of the fluid remaining in the ground is 5,000 to 8,000 feet beneath the surface, lower than groundwater aquifers that may supply drinking water, which are generally no more than 1,000 feet below the surface. But the potential for cracks in cement well casings and the surface escape of chemicals or methane gas during the process of insertion and extraction is real and can pose a threat to groundwater aquifers, as we have seen in the Dimock case discussed earlier. In addition to the fracking chemicals or fluids, the impact of potential methane-gas leaks (potential for explosion and asphyxiation) is a very important concern in relation to ground wells in rural areas. Finally, the potential for errors in waste disposal or improper treatment of the retrieved waste-water are among the other major concerns associated with the relatively unregulated and rapid acceleration of horizontal fracking.[18]

Energy producers are quick to deny that any of the risks associated with fracking represent significant concerns. They reassure us, as a matter of routine but often without the rigorous science to back up their reassurance, that all risks are minimal and manageable. In fact, what we are calling the conflict dynamic colors the conversation about any risks from the very beginning, but the specifics of the dynamic are significantly different from those we discussed in relation to climate change. Like most organized interests and all corporations, energy producers work hard to influence the policy process. They spend great amounts of money to avoid governmental regulation and to, in effect, bury risks associated with their work. In the case of climate change, corporations and fossil fuel interests use a handful of dissident "scientists" to cast doubt on the consensus arrived at by climate scientists. The goal, of course, is to deny the likelihood of any adverse impacts arising from global climate change. The support of faux "experts" who promote research outcomes desired and paid for by the industry is intelligently combined with sophisticated political and public-relations campaigns designed to reduce the visibility of risks and, thus, the likelihood of governmental regulation of the fossil fuel industry. Whether restoring the image of an industry[19] or promoting its interests in avoiding governmental regulation or weakening public awareness of environmental threats posed by their activity,[20] corporations spend immense resources to shape public opinion and influence public policy makers.

In the case of hydraulic fracturing, the conflict dynamic does not work to deny or discredit a scientific consensus that works against its interests; rather, it seeks to prevent an objective scientific assessment of any risks or dangers posed by the new horizontal drilling technologies. Whether facing down a threat to profits in the form of a corporate tax hike or working to prevent any regulatory regime that might restrict their activity, corporations use lobbying and political campaign donations as another way of making money. When it comes to the nation's energy needs, the goal of energy independence, and the number of jobs supposedly supported or impacted by energy production, elected officials are already predisposed to see things exactly as the fossil fuel corporations do. These corporations also spend huge amounts of money to ensure that the voting public sees things their way. There are a few key strategies that corporate lobbying efforts use to bend the government and the public to serve the corporate will. The fossil fuel industry, in particular, uses these strategies most effectively, and they work most of the time. There are eight basic strategies or tactics that might be generally articulated as follows: (1) control the terms of the debate, (2) spin the media, (3) engineer a following, (4) acquire "independent" support, (5) sponsor a "think tank," (6) appear to consult with the public and with opposing groups, (7) neutralize the opposition, and (8) control the web.

First, corporate lobbying seeks to control the terms of the debate. The main goal here is to steer all conversation away from debates they cannot win and on to those that they cannot lose. If a public discussion of a company's negative environmental impact is unwelcome, their lobbyists will push instead to have a debate with politicians and the media on the hypothetical (always exaggerated) economic benefits of their ambitions. Once this narrowly framed conversation becomes dominant, dissenting voices will appear marginal, irrelevant, or out of touch with the important economic realities that matter most.

Step two in the corporate tool kit is to spin the media. The trick is in knowing when to use the press and when to avoid it. Talking to government through the media is crucial. Messages are carefully crafted to have maximum impact on public opinion and the opinion of policy makers. Even if the corporate goal is pure, self-interested profit making, it will always be dressed up to appear synonymous with the wider national interest. For fossil fuel interests, that typically means an emphasis on energy independence, economic growth, and jobs.

A third step is essential for step two to succeed; the corporate lobbying machine needs to engineer a following. Thus, corporate advertisement is a critical component, and its goal is to engineer the creation of a critical mass of voices singing and people dancing to its music. Television advertisements about natural gas and hydraulic fracturing are not meant to sell a product. They are intended to sell a political idea. They are designed to enhance profits by discouraging public policies or regulations that might impose costs on energy producers and thereby reduce profit margins. Their concern is less about the United States' energy future than it is about their future profit margins. A patriotic-sounding message about our national energy needs and a promise of economic miracles if they are allowed and supported to do exactly what they want to do without governmental restriction is designed to hit your "feel good" spot, earn your support, and influence your congress member.

Step four is necessary because corporations are among the least credible sources of information for the public. Corporations, including fossil fuel companies, know this is how they are perceived, so they spend money to acquire authentic, authoritative, and seemingly independent people to carry their message for them. This tactic has been used by tobacco companies in decades past and by climate-change deniers in the present day. Paying scientists or pseudoexperts to produce "evidence" and publicly endorse the corporate position is often an effective way to steer the public discussion exactly where the corporate interest wants it.

Step five flows quite naturally out of step four as it even further enhances the ability to acquire the appearance of authenticity. Sponsoring a think tank is a no-brainer. Think tanks provide companies with a lobbying package that can include generating media-friendly reports, providing "talking heads" for media appearances, publishing policy reports, and influencing politicians. As an added bonus, it may give the corporation a way of being less direct and open in its efforts to influence policy outcomes and lend to the appearance of broader support for its efforts.

A sixth tactic is to "consult" with critics or with the public generally. This might mean running focus groups, exhibitions, planning exercises, and public meetings as a way of flushing out opposition and providing a managed channel through which would-be opponents can voice concerns. Opportunities for critics or the general public to influence corporate objectives or outcomes are almost always nil, but a show of concern about the views of their critics or a response to or a consultation with the public can be useful in creating the conditions necessary for successfully implementing tactic seven—neutralizing opposition.

Corporate lobbyists want government to listen to their message but ignore counterarguments from campaigners, such as environmentalists, who have long been the bane of commercial lobbyists. A range of tools may be employed here, including monitoring opposition groups with online "listening posts" that can pick up the first warning signals of activist activity. Rebuttal campaigns are frequently organized and employed to counter activist activity. There are also more serious activities used primarily when big-money commercial interests are threatened, including infiltrating opposition groups, otherwise known as spying.

In the digital age, tactic eight is very commonly used. It might simply be called "control the web." The way to control information online is to flood the web with positive information or messages about your company.

Lobbying agencies, for example, may create blogs, fake news releases, and various forms of clickbait to create a narrative or define the terms of debate. More negative efforts may include trolling opposition groups and bombarding their chat rooms with contrary messaging.

Corporate efforts to influence public opinion and the decisions of elected office holders are a legitimate part of our democracy, of course. But, as a matter of unfortunate routine, these corporate efforts are often designed to weaken public awareness of environmental threats or public health risks associated with their activities. Indeed, when scientific research documents these threats or risks, corporations often respond by spending millions of dollars on cover-ups, deceptions, data manipulation, fraudulent claims, and fake studies.[21] This often includes assertions that science is not conclusive enough to identify risks or to support regulatory action. At the same time, the assertion that science is not conclusive does not deter natural-gas drillers from being absolute in their insistence that their methods are completely safe. The last thing they want is for science to advance and become more robust in documenting risks. This, one might suggest, is why it is of critical importance that governments play a proactive role in monitoring and regulating for public health and safety. But governments, under the influence of corporate lobbying and public relations campaigns, are often reluctant or tardy with respect to meeting this responsibility. Such was the case as fracking emerged and picked up speed, very much by corporate and political design.

Preemptive Conflict: The Avoidance of Regulation

The 2005 Energy Policy Act passed by Congress (crafted by Vice President Cheney, who once ran Halliburton, one of the companies that pioneered fracking) exempted hydraulic fracturing from meeting the requirements of the Clean Air Act, the Safe Drinking Water Act, and the Clean Water Act.[22] The purpose here, and the goal of the industry and its supporters in government, was to eliminate as completely as possible any meaningful federal oversight of the fracking process. This policy was preceded by a 2004 determination (that was neither comprehensive nor scientifically rigorous) by the Environmental Protection Agency (EPA) that concluded that the extraction of natural gas via horizontal fracking posed little to no threat to drinking water or public health. This study was denounced by at least one EPA whistleblower for its poor science and as having been the product of an industry-influenced review panel.[23] In 2010, the EPA reversed this earlier stance and announced it would launch a $1.9 million research program to assess public health risks associated with fracking.[24] But as this study was conducted over the next several years, it did not slow production in the least.

The EPA study, finally produced in 2015, concluded what was called the most extensive government review of U.S. fracking practices. It found no evidence of widespread damage to drinking-water supplies, but at the same time it warned of the potential for contamination from the controversial techniques used in oil and gas drilling.[25] This report, a bit contorted perhaps, did nothing to cool the national debate over hydraulic fracturing and horizontal drilling. Fracking had spurred huge increases in U.S. oil and gas production. This was viewed positively by the public and policy makers. Speaking out of both sides of its mouth, in this report the EPA sought to warn of some dangers without raining on the natural-gas parade. Opponents and supporters of fracking instantly seized on the portions of the report that supported their view. The national conversation was hardly advanced, and the industry continued to dictate the terms of the debate, or so it would appear. This study was handicapped by the oil industry's refusal to provide key data, and the EPA found, and to some extent ignored, disturbing evidence of fracking polluting our water despite not looking very hard.[26] Clearly, this study struck a balance between cautionary warning and overall support for what the government clearly wanted to do (i.e., support fracking).

The natural-gas industry has consistently claimed that drilling for natural gas has not caused a single case of groundwater contamination. This is not true, of course. From the very beginning of the fracking revolution, the evidence contradicted industry assurances. For example, the Pennsylvania Department of Environmental Protection had documented in Dimock the contamination of an aquifer that filled household wells in a rural area where more than 60 natural-gas wells were drilled in a nine-square-mile area.[27] There were, at the same time, a growing number of other such reports from Pennsylvania to Colorado regarding possible groundwater contamination.[28] Studies from New York asserted that improperly treated fracking wastewater (containing radioactive materials and harmful chemicals) was finding its way into the state's bodies of water.[29]

A very important study concerning the methane contamination of drinking water in conjunction with hydraulic fracturing was published in the spring of 2011.[30] The researchers identified specific concerns, including the toxicity of produced water from fracturing fluids that may be discharged into the environment, fluid and gas flow and discharge into shallow aquifers, the impact on private wells that rely on shallow groundwater for drinking and agricultural use, and the potential for explosion.

They proceeded to conduct tests in Pennsylvania of drinking-water wells in the proximity of fracking activity. Sixty wells were tested, and methane concentrations were found in 51 (85 percent) of them. The average methane concentration in shallow groundwater in active drilling areas was *17 times higher and exceeded the level identified for "urgent hazard mitigation" by the U.S. Office of the Interior.* In this study, there was no evidence of contamination of drinking water by fracking chemicals or fluids, but the correlation of drilling and high methane levels was considered a cause for heightened concern. At a bare minimum, this suggested the need for continued and serious investigation.[31]

The authors of the Pennsylvania methane contamination study recommended long-term monitoring of the industry and private homeowners. They urged drilling firms to comply with a recent request by the EPA to voluntarily report the constituents of fracking fluids. Many companies would ultimately comply with this request for voluntary reporting but only as a means to stall possible regulatory action by the federal government. Most importantly, the study called for systematic and independent data collection on groundwater quality before drilling begins in any region and stressed the need for greater stewardship, more knowledge, and regulation to ensure the sustainable future of shale-gas extraction.[32] In response to this study, the industry resorted to its well-rehearsed script. It said the findings of the study were unconnected to the drilling activity. Thus, they saw no need to conclude that widespread cases of methane contamination were in any way connected to fracking.

Another new study released in the spring of 2011 called into question the notion of natural gas as the cleaner energy alternative. This would appear to cast significant doubts on the benefits of fracking in combating global warming.[33] This study concluded that the greenhouse-gas footprint of natural gas is actually greater than that for conventional gas and oil or coal. How could this be? The study looked at the footprint of shale over a longer time span and included an assessment of waste, leaks, production technology, consumption, and so on. Considering all these things together over the expected lifetime of a well, fracking will contribute more greenhouse-gas emissions than previously thought. Indeed, the overall carbon footprint of shale gas will be 20 percent greater than for coal, according to this analysis.[34] At about the same time this study was published, a number of other studies also suggested that methane has greater global-warming potential than previously assumed.[35] Scientific studies like these challenged the notion that natural gas is the cleaner energy alternative or a bridge fuel to a cleaner energy future in the battle against climate change. Naturally, the natural-gas industry immediately questioned the accuracy of all such studies. Upon the release of any new research on the risks or negative environmental impacts associated with fracking, the industry is always quick to react by refuting the legitimacy of the conclusions or the methods of analysis as it steadfastly reaffirms the safety of its drilling technology. Whatever scientific inquiry may show as new studies are produced, the industry always says that science is inconclusive and the risks unproven. The questions raised, however, are serious. They demand further study and rigorous scientific research. That, of course, is the last thing the industry really wants.

The oil and gas industry sponsors and spins research to shape the scientific debate over horizontal hydraulic fracturing. In turn, the oil and gas industry alleges that fracking opponents are responsible for their own share of deceit as they try to align research with their arguments. It can be very difficult for the policy maker and the public to sort out the truth. One wonders whether politicians want the truth or if they are only interested in the science that supports what they already believe or the things they have already decided to do. The industry, of course, wants to preemptively discredit any and all scientific research that suggests environmental or health risks associated with their activity. This is the driving force of the conflict between science and politics over this issue. The influence of the industry on the dialogue and its support for sympathetic conservative policy makers who are already averse to things like governmental regulation and environmentalism combine to limit the conversation.

An example of how the preemptive narrowing of the conversation works can be seen in the evaluation of groundwater contamination. The EPA knew about groundwater contamination from fracking as far back as 1987. In fact, links between shale drilling, fracking, and groundwater contamination have been well documented. In Pennsylvania, it was found that the closer you live to a well that hydraulically fractured underground shale for natural gas, the more likely it is that your drinking water is contaminated with methane.[36] This was the conclusion of a study published in the Proceedings of the National Academy of Sciences of the United States of America in July 2013. It was a first step in determining whether fracking in the Marcellus Shale underlying much of Pennsylvania is responsible for tainted drinking water in that region. The study found methane in 115 of 141 shallow residential drinking-water wells. The methane concentration in homes less than one mile from a fracking well was six times higher than the concentration in homes farther away.[37]

Most groundwater supplies are only a few hundred feet deep, but if the protective metal casing and concrete around a fracking well leak, methane can escape into them. The study did not prove that fracking has contaminated specific drinking-water wells, however. The drilling companies have persistently denied even the possibility that some wells may leak and that this may contaminate water. A study in 2017 reported a case where natural gas and other contaminants migrated laterally through kilometers of rock at shallow to intermediate depths, impacting an aquifer used as a potable water source. This study represented the first peer-reviewed paper *conclusively confirming that fracking can and does contaminate drinking water supplies.*[38] As always, the industry response was to spend time and money to keep policy makers and the public in the dark on the real impacts of fracking, including funding front groups to obscure the science that assesses the risks associated with fracking and drilling.

Hydraulic fracturing has been tied to environmental risks such as spills for a long time. At a minimum, there is sufficient evidence to demonstrate that we should have some urgency about researching and documenting any resulting environmental and health risks. But the preemptive battle waged by the industry has significantly delayed that analysis. Their overriding goal has been to prevent any regulatory or policy restrictions that might impact their profit margins. Congress has been tamed by these industry efforts, as the appetite for policy interventions to regulate fracking has practically disappeared. Presidents have proven to be lovers of natural gas as well, and there is little serious questioning of horizontal drilling or any of the possible negative by-products associated with it. Even the Environmental Protection Agency has been controlled by the industry's preemptive strike. As we have seen, the EPA's conclusions regarding groundwater contamination were a contorted contradiction. The EPA found no evidence of widespread damage to drinking-water supplies, but at the same time it warned of the potential for contamination. On the subject of chemical spills, the EPA has for some time severely underestimated the number of these occurrences. In its 2015 report, the EPA said "the number of spills nationally could range from approximately 100 to 3,700 spills annually, assuming 25,000 to 30,000 new wells are fractured per year."[39] The EPA reported only 457 spills related to fracking in 11 states between 2006 and 2012. Of course, we would soon discover that these numbers were off by a very wide margin.[40]

A 2017 study demonstrated that fracking-related spills occur at a much higher rate than either the industry admitted or the EPA reported. The study, published in the journal *Environmental Science & Technology*, revealed 6,648 spills in four states alone—Colorado, New Mexico, North Dakota, and Pennsylvania—in 10 years.[41] This research confirmed that 16 percent of fracked oil and gas wells spill hydrocarbons, chemical-laden water, fracking fluids, and other potentially dangerous substances. State-level spill data associated with unconventional oil and gas development at 31,481 fracked wells in the four states between 2005 and 2014 was carefully examined. The 6,648 spills over the 10-year period was the equivalent of 55 spills per 1,000 wells in any given year. Why did the researchers' numbers for just four states vastly exceed the 457 spills (in 11 states) estimated by the EPA? This was primarily because the agency only accounted for spills during the narrowly defined hydraulic fracturing stage (i.e., the fracturing of the rock) itself rather than the entire process of unconventional oil and gas production. This worked to the advantage of the natural-gas industry but not so much the national interest in safety and health. It is important to understand and assess spills at all stages of well development. For example, hydraulic fracturing requires the transport of chemicals and materials to and from the well site. Additionally, the storage of these materials on site needs to be included as an area of concern. The study found that 50 percent of spills were related to storage and moving fluids via pipelines.[42]

Perhaps one of the most unanticipated by-products of the hydraulic fracturing process is an increased risk of earthquakes. It is not actually fracking but the injection of wastewater into disposal wells that seems to be the problem. Fracking involves, as we have seen, the high-pressure injection of fluids (water and chemicals) and sand into the ground to open up cracks in shale rock so that natural gas may escape and be captured. This, according to some recent studies, lowers the barriers to earthquakes by loosening the rocks enough to make an earthquake more likely. In Arkansas, it was discovered that wastewater disposal wells (i.e., high-pressure injection of wastewater into dead wells) were associated with an increase in earthquake activity. The number of incidents diminished quickly back to normal levels when wastewater wells were shut down.[43]

The U.S. Geological Survey has concluded that fracking is not the cause of most induced earthquakes. The earthquakes occurred, according to this survey, as a result of wastewater disposal.[44] But to say fracking is *not* causing most of the induced earthquakes, while technically true, is not honest. Wastewater disposal is part of the fracking process. As the primary cause of the recent increase in earthquakes in fracking communities, it is hard to say that fracking is irrelevant to the issue. Between 1970 and 2000, there was an average of 20 earthquakes per year within the central and eastern United States. Between 2010 and 2013, there was an average of more than 100 earthquakes annually. The U.S. Geological Survey agreed that fracking wastewater, a by-product of oil and gas production that is routinely disposed of by injection into wells specifically designed for this purpose, is linked to this dramatic increase in earth-quake activity.[45]

In fairness, there is a need for subsequent study of the risks associated with fracking. In some areas we are just beginning the analysis, while in others the research raises extreme concerns that need to be taken seriously. What is more, the cumulative impact of fracking activities over the next 20 or 50 years may dramatically alter any of our current notions about hazard risks associated with horizontal fracking. One would suspect, based on the pattern of the findings to date, that these risks will be greater than we presently know. The continued study and refinement of our understanding of both the immediate and cumulative impacts of fracking on the environment, the water supply, the air, human health, and the safety of our communities should be regarded as an urgent necessity and not an inconvenience imposed by people or groups opposed to fracking. It is, rather, a vital component in the smart and necessary forward thinking and planning that is required to create and maintain sustainable and resilient communities. But our policy makers do not seem to be interested in this kind of thinking.

There have been efforts at the federal level to legislate and address the concerns associated with hydraulic fracturing. In 2009, almost a full decade into the fracking revolution, Congress tried to address these concerns in a serious way. This effort failed but was renewed in 2011 with the reintroduction of the Fracturing Responsibility and Awareness of Chemicals Act (FRAC Act). The Senate version of this bill would have closed the oversight gap that the natural-gas industry has benefited from since the passage of the 2005 Energy Policy Act and repealed the provision of the 2005 act that exempted the industry from complying with the Safe Drinking Water Act. The bill would also have required the public disclosure of chemicals used by the natural-gas industry in its fracking operations, although the companies would not be required to reveal specific formulas where there is a proprietary interest. However, a provision required that proprietary chemical formulas be released to attending physicians, the state, and the EPA where necessary for treatment in emergency situations.[46] But even these simple steps were impossible to take.

The FRAC Act died in committee; in fact, it died twice. The natural-gas industry, as one would expect, lobbied aggressively against this legislation as part of its overall agenda to limit federal oversight of gas drilling. One of the industry arguments was that state regulation was superior to federal regulation because it would be fine-tuned to local conditions best assessed by local officials. Of course, a patchwork of state regulations worked to the advantage of the industry. This was especially true in the energy-rich but job-poor states where much of the fracking occurs. Friendly legislators in these states could be counted on to tread softly where state regulation was concerned. The U.S. Congress was and has remained far from united in perceiving the need to act. It is clear that the process of risk assessment and risk management will most likely continue to be pursued in the U.S. policy process in an overtly partisan manner, and, as such, it will continue to be far from efficient in serving the public interest in safety and health. A responsible approach to risk assessment and risk management is perhaps compromised to the degree that it is dependent on the vagaries of partisan politics and economic self-interest. One cannot help but wonder if there is any potential for responsible action.

Whether natural-gas drillers, members of Congress, or the general public will ever take seriously the threats posed by hydraulic fracturing remains to be seen. It is interesting to note that one major insurance company has taken them very seriously. In the summer of 2012, Nationwide Mutual Insurance Company became the first to publicly state that it would not cover damages caused by hydraulic fracturing. Nationwide said that its policies were not designed to cover the risks posed by fracking, which were too great to ignore and too risky for them to even consider covering.[47] Advances in hydraulic fracturing have put trillions of dollars' worth of previously unreachable oil and natural gas within humanity's grasp. While it might be perceived as a good thing in relation to the nation's energy needs, the question remains: Have we done an adequate job of assessing the costs and benefits of fracking?

In 2014, a Stanford-led study assessed the environmental risks associated with hydraulic fracturing and concluded that we needed to do much better in understanding and managing these risks.[48] The study addressed not only greenhouse-gas impacts but also fracking's influence on local air pollution, earthquakes, and, especially, supplies of clean water. The Stanford study assessed the positive (e.g., cleaner burning) and negative (e.g., groundwater contamination) impacts associated with hydraulic fracturing for natural gas, and both the costs and benefits of fracking were presented in a balanced fashion. The study also highlighted several policies and practices that could optimize fracking's environmental cost-benefit balance. Most importantly, it highlighted the need for further research in several important areas. For example, the direct impact on the health of residents living near drilling sites is virtually unexplored.[49] The point to be taken from this study is that the fracking revolution of the 21st century has gone ahead at laser speed without anything approaching an adequate assessment of environmental and health impacts.

Europeans are displaying perhaps a bit more urgency than Americans about assessing the environmental and health risks associated with fracking. At least European governments are. They emphasize that hydraulic fracturing has the potential to cause fugitive methane emissions, air emissions, water contamination, and noise pollution. Water and air pollution are identified by most of the EU as the biggest risks to human health from fracking. There is an insistence in European governmental circles on research to determine if human health has been affected. Likewise, there appears to be greater urgency than in the United States about the need to develop and adhere to serious regulations and safety procedures that are seen as necessary to avoid negative impacts.[30]

The American people are actually beginning to show more concern about the risks associated with hydraulic fracturing than their government has shown to date. After supporting fracking for natural gas throughout most of the first decade of the new century, the public is now moving in the other direction. In 2016, a Gallup survey showed public opposition to fracking mounting in the United States (see Table 4.1).

Fracking has become a contentious topic in American life. It has been seen as a source of great prosperity for the nation's crude-oil producers. Politicians have supported it as a source of plentiful, cheap, and "cleaner" energy. Also, because of the economic implications, especially where jobs are supposedly concerned, policy makers have had very few concerns about natural-gas development. As is the case with virtually all matters of substance, American public opinion seems divided very much along partisan lines. Republicans favor fracking and oppose any regulation of it at much higher levels than do Democrats and independents. Of course, as the public becomes more aware of some of the possible negative impacts associated with the process of hydraulic fracturing, it might be expected

Table 4.1   Do You Favor or Do You Oppose the Practice of Hydraulic Fracturing?

	Favor	Oppose	No Opinion
2016	36%	51%	13%
2015	40%	40%	19%

	2015 Favor	2016 Favor
Republican	66%	55%
Independent	35%	34%
Democrat	26%	25%

Source: Gallup, http://www.gallup.com/poll/190355/opposition-fracking-mounts.aspx

that the number of those opposed to fracking might increase. Becoming more aware is a process that has only just begun.

In 2016, the U.S. Geological Survey published a report stating that 7 million Americans could experience man-made earthquakes as a result of fracking activity.[51] The part of the United States that has been most affected seems to be Oklahoma, which the report says has a one in eight chance of experiencing damaging quakes. In fact, the third-strongest earthquake recorded in Oklahoma history (5.1 magnitude) occurred in February 2016. In 2015, Oklahoma had over 900 quakes that were a 3.0 magnitude or more. Kansas, Texas, New Mexico, Arkansas, and Colorado are the other states where fracking-related quakes are most frequently experienced.[52] While public opinion is beginning to skew against the fracking process, policy makers might best be described as polarized. Predictably, the profracking policy makers agree with the fossil fuel industry and lobbyists, while the antifracking or the proregulation policy makers generally agree with environmental groups while generally disagreeing with the industry. The antifracking or proregulation group sees water contamination as a major issue, while the profracking group does not see it as a significant problem at all. The profracking policy maker strongly agrees that fracking provides benefits to the nation's economy that far exceed any concerns about the environment or the safety of the process. The antifracking policy maker strongly supports, at a minimum, the need to regulate the process for safety.

It is entirely possible that the hydraulic fracturing debate is, like all our political debates, unnecessarily polarized. But it is also true that the natural-gas industry is unwilling to acknowledge any problems whatsoever. The unwillingness of the industry and the politicians who support them to take the risks seriously and their preemptive campaign to deny and to obstruct any efforts to regulate their activities in the interest of public safety complicates research efforts by preventing the release of some data. This inevitably makes it look like the industry is hiding something. The industry and the political right have consistently said that there is little if any credible scientific evidence of fracking's feared harms. They continue to insist that the overwhelming scientific evidence of its environmental benefits, including substantial reductions in both local and global pollutants, outweigh any need for concern. In previous years, this position may have been defensible as the science was still pretty ambiguous and had not really sunk its teeth into the assessment of risk, and a great deal turned on how narrowly "fracking" was defined. Technically, "fracking" only refers to the water and chemical blast, not the drilling, the disposal of waste, or the huge industrial operations that accompany it all. But when one considers the fracking process to include all these phases, the risks are greatly multiplied.

It is entirely possible that now, nearly two decades into the fracking revolution, we are only beginning to do the science that will provide the insight that we and our policy makers truly need. But the conflict dynamic that has preempted the science and enabled fracking to leap out far ahead of it may be escalated to science denial by the industry as the necessary strategy to defend its interests. That seems to be how the game is played as science and politics collide. As we conclude this discussion about fracking, it is important to understand the unique preemptive nature of the conflict dynamic that has been at work since the beginning of the fracking revolution and to anticipate how the dynamic will require the industry's strategy to shift as the science advances. In essence, the conflict dynamic will continue, much to the detriment of the public interest.

Conclusion

It can be said that scientists have a reverence for evidence. For policy makers or politicians, scientific evidence is at best an afterthought. Therein is the root of an inevitable conflict between science and politics. While scientists produce evidence that supports an objective foundation for fact or truth, politicians frequently make ideological assertions not supported by any fact. Inevitably, as the contest between scientific fact and ideological assertion slowly plays itself out, the public interest is ill served. Industry, in the meantime, is interested only in the science that will create and support the technologies and engineering systems that will expedite its work and maximize its profits. Any science that identifies the need for industry to bear the costs of any risks will initiate a prolonged conflict between science and politics, as industry and its supporters in the policy-making process, armed with "alternate facts" and their own "science," will stall the process of identifying and addressing the risks. Where public health and safety is at stake, this is a loss for everybody.

There are a host of environmental, health, and safety hazards associated with hydraulic fracturing for natural gas that continue to make fracking a hot-button issue that divides Americans. Whatever the benefits associated with fracking, there are a number of very dangerous things that can happen when you drill a hole in the surface of the earth, inject toxic chemicals into that hole at high pressure, and then inject the wastewater deep underground. It would make very good sense to understand those dangerous things, to calculate their probability, and to take preventive action to regulate the process to avoid these dangerous things where possible. Indeed, it would seem to be a very good idea for policy makers to consult with scientists, obtain the best scientific assessment of these risks, and move efficiently to manage them.

The basic numbers associated with fracking suggest the wisdom of risk assessment and risk management. Consider that, as of 2015, there were over 1.1 million active gas wells in the United States. The number continues to grow, but let's look at what 1.1 million active wells meant with respect to other numbers. It takes about 72 trillion gallons of water and about 360 billion gallons of chemicals to run 1.1 million wells. It takes 40,000 gallons of chemicals and 8 million gallons of water for each fracking site. The number of chemicals used, at least potentially, in fracking fluids is 600. These fluids include known carcinogens and toxins, such as lead, benzene, uranium, radium, methanol, mercury, hydrochloric acid, ethylene glycol, and formaldehyde.[53]

While the research is ongoing and some conclusions are tentative, science has advanced enough to document potential health risks in communities most directly impacted by fracking activity. According to the Natural Resources Defense Council, enough is known to say that there are serious health impacts associated with fracking, including respiratory problems, birth defects, blood disorders, cancer, and nervous system impacts. Enough is already known to raise serious concerns for workers and people living closest to wells as well as entire regions with high volumes of oil and gas activity.[54] An emerging pattern in the science reveals unsafe levels of air pollution near fracking sites around the country. Much more research is needed to better understand these threats and a wide range of others that have emerged.

The 72 trillion gallons of water required to run over a million wells is a concern unto itself. This directly reduces the amount of clean water available to surrounding residents. Also, when water is not available to fracking sites locally, it must be transported from other regions. This ultimately draws down available water from lakes and rivers across the country. Water contamination could also reduce the overall water supply of regional fracking areas. As we have seen, the chemicals that are used in the process have the propensity to leak back into local water supplies. Wastewater is also an issue at fracking sites. The water used for fracking that is returned to the ground surface consists of toxic contaminants. Of course, with chronic drought conditions in much of the country, the use of so much water for fracking may contribute to major water shortages. The redirection of water supplies to the fracking industry not only causes water price spikes but also reduces water availability for crop irrigation.[55]

Methane is a main component of natural gas. It is about 25 times more potent in trapping heat in the atmosphere than carbon dioxide. As we have seen, surface leaks and leaking wells are a major concern associated with fracking. In addition to things like the contamination of water supplies, what this might mean for air quality and in relationship to climate change are concerns of some interest. The number of other air contaminants released through the various drilling procedures, including construction and operation of the well site, transport of the materials and equipment, and disposal of the waste, are reasons for serious concern. In fact, just a list of "concerns" associated with fracking is impressive at a glance. These include contamination of groundwater, methane pollution and its impact on climate change, air pollution impacts, exposure to toxic chemicals, waste disposal, blowouts due to gas explosion, large volume of water used in water-deficient regions, fracking-induced earthquakes, and infrastructure degradation.

Whatever one might ultimately think about fracking, the concerns raised by this practice are very serious, and they need to be addressed in the interest of public safety and health. Some people, and even some environmental groups, believe natural gas—and thus, fracking—serves as a pivotal bridge in the transition from reliance on dirty fossil fuels like coal and oil to a future of clean energy. In other words, it buys time for the production of solar, wind, and water power. Others, including other environmental advocates, say dependence on natural gas from fracking is merely a lesser evil. It is not the solution to anything. Still others simply say we need to go full speed ahead with fracking to meet our national energy needs and to bolster our economy. Many in this last group might say environmental and health concerns are liberal deceptions, so drill, baby, drill! Since the beginning of the fracking revolution, there has been a never-ending tug-of-war between industry and environmentalists over how much regulation of the industry is needed. The industry has, thus far, won that tug-of-war by a long shot. Before science even really entered the game, this is what the fracking playing field looked like, and these were the teams.

As science has begun to engage some of the concerns surrounding the fracking process, and as its findings have become more certain in some areas while demonstrating the need for much more analysis in others, its work has been interpreted by the warring parties in the context of their prevailing assumptions and goals. The industry tends to deny risks or be indifferent to them until a disaster happens. Keeping production costs down, avoiding costly or inconvenient regulation, and maximizing profit are invariably more important goals in the scheme of things. Sacrifices in safety and environmental disregard are all too commonly incentivized by management to maximize profit, and a rolling of the dice on risk is simply business as usual. For the politician, ideology is preferable to knowledge in approaching an issue and waging a debate. The same might be said for almost any interest group seeking to influence the policy process. Science is, if it is anything at all, an afterthought for industry, policy makers, and political actors.

If science had been seen as an important partner in the fracking revolution, how much differently would the revolution have proceeded? Would it have proceeded? Assuming that it would have, it would have played out much differently than it did. The EPA research program to assess the environmental and hazard risks associated with fracking would have been pursued as a first step in the policy process and completed, or at least advanced, years earlier. The exemption of natural-gas producers from federal regulation (i.e., the Safe Drinking Water Act, the Clean Water Act, and the Clean Air Act) would not have been enacted prior to the scientific study to inform the policy. The void created by removing the EPA as a monitor on the boom in gas fracking—a boom that has cut across state lines into almost every region of the country with the potential to impact public health, safety, and sustainable hazard resilience across the country—would not have been allowed to exist. The legitimate concerns related to the contamination of drinking water, the toxicity of fracking chemicals and wastewaters, the correlation of drilling and high methane levels in ground wells, and a host of other concerns would have been much more thoroughly explored much earlier in the process. Responsible interim legislation, perhaps something like that proposed in the FRAC Act, would have been enacted years ago. By now, almost two decades into the fracking boom, the foundation for knowledge-based decision making would have led to the establishment of sensible federal and state regulations protecting public health, mitigating hazards, promoting safety, and ensuring, no doubt, the sustainable future of shale-gas extraction. Policy makers and the natural-gas industry would, through the determined application of scientific research, have from the very beginning been full partners and collaborators with science in the tasks of risk identification, vulnerability assessment, and mitigation.

It is also possible that a more collaborative relationship between science and politics would have led to the decision to back off fracking, and fossil fuels in general, and prioritize the acceleration of the development and production of renewables in response to climate change. Of course, that would be to assume that the conflict dynamic that has dominated the discussion about climate change doesn't exist. But it does. And what we have seen in relationship to natural-gas fracking is simply a continuation of that dynamic. The fact is that the science that might have guided political and industry leaders to make intelligent decisions that reflect the public's best interest was not seen as indispensable or even very important. It was not regarded as even the least bit necessary. Under these conditions, when the science does begin to emerge or catch up, it gets plugged into the ongoing political battle and is not respected. Science, like facts in today's political arena, is optional, and everybody is entitled to their own.

CHAPTER FIVE

Evolution and Creationism: A Classic Case of Resistance

Preachers had set up their revival tents along the main street of the small city. The faithful were stirred up, but they weren't the only ones. Banners lined the streets, and a thousand people had shown up at the local court-house. Chimpanzees had been brought into town. They were thought by some to be there to testify at the trial about to begin, but actually they were there to perform in a sideshow extravaganza on Main Street. Ven-dors sold Bibles, toy monkeys, lemonade, and hot dogs. This was all a part of the carnival atmosphere that prevailed on July 10, 1925, in the small community of Dayton, Tennessee, at the beginning of the so-called Mon-key Trial.[1]

The state legislature in Tennessee had enacted a law in March 1925 that made it a misdemeanor, punishable by a fine, to "teach any theory that denies the story of the Divine Creation of man as taught in the Bible, and to teach instead that man has descended from a lower order of ani-mals."[2] A local businessperson in Dayton, offended by this legislative act, conspired with others, including a young high school science teacher named John Thomas Scopes, to challenge the law in court. It was arranged for Scopes to be "caught" teaching evolution, whereupon he was duly charged in violation of Tennessee state law.[3] The trial became a national sensation. William Jennings Bryan, three times chosen and three times defeated as the Democratic nominee for president, volunteered to assist the prosecution. The American Civil Liberties Union, joined by the great defense attorney Clarence Darrow, organized the defense for what became one of the most famous trials in American history.

Most Americans know a little bit about the Monkey Trial. Darrow and the defense sought to attack the Tennessee law banning the teaching of evolution as unconstitutional. Bryan and the prosecution sought to portray this as a coordinated attack on Christian fundamentalism. Darrow expected to lose at trial but hoped for an appellate decision that would declare the law to be unconstitutional. Indeed, the judge at the trial, who began each session with a prayer, disallowed the introduction of any scientific testimony by experts on the grounds that the law was not in question. The only thing to be decided was whether or not Scopes had violated the law. Darrow, changing his tactics, then called Bryan as his sole witness in an attempt to discredit his literal interpretation of the Bible. In what many felt to be a win for the defense, Bryan was subjected to severe ridicule as Darrow more or less forced him to make ridiculous and contradictory statements to the amusement of the crowd. Indeed, Bryan's reputation did suffer as a result. But knowing the trial was lost before it began, Darrow closed by asking the jury to return a verdict of guilty against his client so that the case might be appealed under Tennessee law.[4]

After eight minutes of deliberation, the jury returned with a guilty verdict. The judge ordered Scopes to pay a fine of $100, the minimum the law allowed. Bryan had won the case, but he had been publicly humiliated, and his fundamentalist beliefs had been disgraced. That's how the national media portrayed the events of the trial. Five days later, on July 26, and still in Dayton, Bryan returned to his room to lie down for a nap after a hearty dinner. He never woke up again. In 1927, the Tennessee Supreme Court overturned the Monkey Trial verdict but not on constitutional grounds. Rather, it based its ruling on a technicality and left the constitutional issues unresolved. They would remain unresolved until 1968, when the U.S. Supreme Court overturned a similar Arkansas law on the grounds that it violated the First Amendment.[5]

Charles Darwin's *On the Origin of Species* was published in 1859. Historians have argued that the Scopes trial was one of the most dramatic events that came about in the wake of Darwin's landmark book.[6] It was, one might say, an inevitable event. Darwin's theory of evolution sent shock waves across the world. Scientists and naturalists embraced it, but some people found it more than a little disturbing. In the United States, churchgoers and religious leaders debated whether to accept modern scientific theory, especially as it pertained to the origins of humans, or to reject it in favor of their traditional literal reading of scripture. Many urban churches decided to reconcile evolution with their beliefs, but rural churches maintained a stricter stance and a more literalist interpretation of the Bible.

The question of who will dominate American culture is ever present. The 1920s was a period of particularly intense conflict between the modernists and the traditionalists of that age. In that context, the guilt or innocence of John Scopes, and even the constitutionality of Tennessee's antievolution statute, mattered little. The meaning of the Scopes trial can best be understood through its historic place in the inevitable conflict between social and intellectual values.[7] In the United States, this conflict seems almost never to be resolved. In fact, it grows more and more intense as new scientific advances and discoveries appear to be in conflict with long-held beliefs and values. In many European and Asian cultures that are technologically and scientifically advanced, this conflict in the United States is perplexing. How, they ask, can Americans living in a scientifically advanced country and embracing a modern lifestyle still debate the reality of evolution?

It is often the case that there is a tug-of-war between two opposite but ever-present and observable traits in our human nature. If science is interesting, we appreciate it and are often fascinated by it. But if science comes into conflict with our deeply held beliefs, we are quite prepared to reject it. That is a story line that runs through the whole of human history. Galileo, for example, could not even persuade leaders of the Catholic Church to look through his telescope as they condemned his teachings that the earth traveled around the sun. Contradicting the teachings of the church was simply not something one could do in the 17th century. But rejecting science in the name of religion or cultural beliefs is not simply a relic of our past or a distant memory; astonishingly, it continues in the United States even into the 21st century. Knowing why it still happens and what it costs us is critical to our understanding of the collision between science and politics. There is perhaps no better place to begin cultivating that understanding than in the contemporary iteration of the evolution-creationism controversy. What is old in this clash is new again, or so it forever seems.

Explaining the Resistance to Science

Let us recall what we said about the resistance dynamic in chapter 1. We noted that this dynamic applied to situations where strongly held cultural values are in opposition to the findings or conclusions of science. Significant segments of the public, we said, may be opposed to science on religious grounds, for example, or for reasons of ideology. Now many will criticize the theory of evolution because it is, well, "just a theory." This is to completely misunderstand what a scientific theory is. It is not a mere hypothesis or just a guess. A scientific theory is an explanation that is confirmed by all the validated experiments and known observations to date. Scientists always recognize that there is a chance, however small, that some of their inductive reasoning is wrong. That's why they keep making observations and collecting new data, even when a scientific theory has been validated. But the theory of evolution is not wrong. Put differently, as many have noted, the statistical odds of it being wrong are less than the odds of the earth being destroyed by a meteor within the next five minutes. The theory of evolution is the closest thing to an indisputable fact that science can discover. Nevertheless, there are still a good number of Americans who do not accept or believe in the theory of evolution (see Table 5.1).

According to a 2014 Gallup survey, more than 4 in 10 21st-century Americans continue to believe that God created humans in their present form less than 10,000 years ago. This is a view that has changed very little over the past three decades. The percentage of the U.S. population choosing the creationist perspective as closest to their own view has fluctuated in a narrow range between 40 and 47 percent since the question was first asked by Gallup over three decades ago. There is little indication of any sustained decline in the proportion of the U.S. population who hold a creationist view of human origins. This means that 40 to 47 percent of Americans have persistently said that the creationist explanation for the origin of human life best fits their personal views. According to the Gallup surveys over this period, these creationist Americans tend to be highly religious. The results consistently underscore the degree to which many Americans view the world around them through the lens of their religious beliefs. The good news in the 2014 Gallup survey (Table 5.1) is that, while only 19 percent believe in evolution minus divine intervention, another 31 percent accept evolution as a scientific reality while still seeing the guiding hand of God in the process. That is to say that half of the

Table 5.1   2014 Gallup Creationism/Evolution Survey

Which of the following statements comes closest to your view on the origin and development of human beings?	
God created humans in present form	42%
Humans evolved, with God guiding	31%
Humans evolved, God had no part in the process	19%
No opinion	8%

Source: Gallup, http://www.gallup.com/poll/170822/believe-creationist-view-human -origins.aspx

population basically accepts the science regarding evolution. What about the 40 or so percent who do not? Those who adopt the creationist view were found to have lower education levels. At the same time, given the strong influence of religious beliefs, it is not clear to what degree having more education or different types of education might affect their views. In other words, highly educated but extremely religious respondents identify as creationists also.[8]

The ongoing discontinuity between the beliefs that many Americans have about evolution and the general scientific consensus regarding it continues to have political and policy implications. At the beginning of 2017, 92 years after the Scopes trial, a number of state legislatures introduced a host of new antievolution and antiscience bills. On March 6, 2017, Arkansas's House Bill 2050 was introduced. This bill would, if enacted, "allow public schools to teach creationism and intelligent design as theories alongside the theory of evolution."[9] Iowa legislators introduced and referred a bill to the House Education Committee on March 1, 2017, that would, if enacted, require teachers in Iowa's public schools to include "opposing points of view or beliefs" to accompany any instruction relating to evolution, the origins of life, global warming, or human cloning.[10] Alabama's House Joint Resolution 78 introduced and referred to the House Rules Committee on February 23, 2017, would, if adopted, urge state and local education authorities to promote the academic freedom of science teachers in the state's public schools. "Biological evolution, the chemical origins of life, global warming, and human cloning" are specifically identified as controversial subjects where "opposing views" should be explored.[11] A pair of bills introduced in the Florida legislature—House Bill 989 and Senate Bill 1210—are ostensibly aimed at empowering taxpayers to object to the use of specific instructional materials in public schools. Materials may be removed on the grounds that they fail to provide "a non-inflammatory, objective, and balanced viewpoint on issues." There is ample reason to believe that evolution and climate change are among the main targets of this proposed legislation.[12]

Neither the bills proposed by state legislatures in 2017 nor any similar proposals passed and enacted in recent years in states like Tennessee and Louisiana are aimed at preventing the teaching of evolution. That was long ago declared unconstitutional. In 1968, the U.S. Supreme Court ruled in *Epperson v. Arkansas*[13] that a state law allowing the teaching of creation, while disallowing the teaching of evolution, advanced a religion and, as such, was a violation of the Establishment Clause of the U.S. Constitution. Creationists then started proposing and passing laws that required teachers to teach evolution as an unresolved "controversy," granting equal time to creationism. This tactic was also struck down by the Supreme Court in 1987 in *Edwards v. Aguillard*.[14] Many of the state laws proposed today are designed to allow teaching things contrary to science on the grounds it promotes critical thinking. Promoters and defenders of this approach decry the "censorship" of nonscientific ideas and advocate allowing teachers to teach "both sides" of scientific theories. This is essentially the preferred method for conservative groups to circumvent a federal ban on the teaching of creationism in public schools. Why, one may ask, almost a century after the Monkey Trial, do we see state after state still introducing such measures? Is it simply a matter of religion versus secularism?

It is true that religion played a role in the immediate 19th-century reaction to Darwin's landmark 1859 publication, but it did not play the dominant role that it does today in the discussion of evolution in the United States. It was in the aftermath of World War I that religion came to dominate the public discussion of the topic. The devastation of and the challenging uncertainties facing the nation after the Great War, combined with the growing animosity of many American Protestants to historical or literary criticism of the Bible, produced a popular reaction against modernism. Fear of what was seen by many as a dangerous and deliberate departure from tradition and values once held to be beyond question, fear of the pace of change in the world generally, and a discomforting uncertainty about the future combined to create a reaction against newer forms of thinking and a desire to restore traditional values. It was in this context in the early 20th century that Christian fundamentalism as we know it today emerged. It regarded evolution as an attack on the Christian faith. Political animosity toward evolution, as we continue to experience it, is ultimately the product of the religious animosity toward evolution as it emerged in the early 20th century.[15] Well into the 21st century, this animosity is still alive and well and continues to shape the dialogue in many respects.

By the second half of the 20th century, ongoing studies and a convergence of evidence from diverse fields had concluded with unmistakable clarity that evolution was a scientific fact. Evolution had taken a central position in the teaching of biology by that time. Professional, scientific, and governmental organizations partnered to provide systematic curricula to help students develop the needed analytical skills to understand science, compete in a global scientific community, and maintain American leadership advantages during the Cold War. This was a time when dominance in science was considered to be necessary for political dominance on a global scale. As we have seen, by the end of the 20th century, state antievolution laws were stricken by the U.S. Supreme Court. But even as this was happening, the fundamentalist opponents to evolution defended their resistance to evolution with a renewed vengeance. Seeing it as an attack against their worldview, knowing they could not ban the teaching of evolution, they created a number of new responses to continue the war against evolution.[16]

Among the strategies employed by those opposed to evolution, as we have already seen, was to demand that evolution be treated as a controversy, that creationism be taught alongside evolution, that alternative "scientific" views critical of evolution be taught, and that alternative explanations or theories be required along with evolution. As we have seen, subsequent court battles surrounding these efforts have unfolded over the years. Most of the efforts have failed to pass constitutional muster. But it must be remembered that each legal case is but a moment in a battle that has been going on for a century, and, as new state laws continue to be proposed, that battle promises to continue for decades to come.

One of the most creative efforts to smuggle creationism into the classroom was the so-called intelligent design movement. The argument here was that a careful study of nature revealed "evidence" of intelligent design. Intelligent design was presented as a science and was said to prove that the appearance of complexity in nature categorically cannot be explained through natural causes alone. In other words, the observed complexity is said to be of such a level that it could not possibly be explained as a natural development. It requires the guidance of an intelligent agent. While not saying who (or what) the intelligent designer was, this "science" was favored by creationists and was seen as a means to get the Christian God back into the classroom discussion.[17] Of course, it might also have gotten technologically advanced aliens into the classroom discussion. One suspects that this would not have been a development welcomed by creationists. In any event, a district court ruled in 2005 that intelligent design was essentially a religion and, as was the case with creationism, could not be included as part of the science curriculum in public schools.[18] While this battle to rescue God from science has consumed creationists, it is important to note that not all religions or religious people found it necessary to deny evolution.

It is instructive to consider at this point the interesting and very different route one religion has taken with respect to the question of evolution. As we have previously noted, and as Galileo could tell you were he still around, the relationship between science and the Catholic Church has been a wee bit testy at times. Ironically, the Catholic Church has largely sat out the cultural battle over the teaching of evolution. This may in part be because the well-established system of Catholic schools means that state laws relating to the public-school curriculum are of much less concern to Catholic clergy. But it is also due to the fact that the church has never really seen evolution as contradictory to its teachings. By the beginning of the 20th century, the Catholic Church had taken a much more flexible approach to the interpretation of the Genesis account of creation than do some of the more literalist Protestant denominations.[19] Two 20th-century popes in particular, Pius XII and John Paul II, clarified the Catholic position very nicely.

Pope Pius XII was a very conservative pope and a very conservative man. He would certainly not be considered to be a man of science. It can be said that he believed or hoped that evolution would prove to be nothing more than a passing scientific fad, but in his 1950 encyclical *Humani Generis*, he stated that nothing in Catholic doctrine is contradicted by a theory that suggests one species might evolve into another. "The Teaching Authority of the Church does not forbid" research, discussion, or teaching "with regard to the doctrine of evolution, in as far as it inquires into the origin of the human body as coming from pre-existent and living matter." The Catholic faith "obliges us to hold that souls are created by God," but it does not oblige us to deny the process of evolution.[20] Pius XII did caution, however, that he considered the jury still out on the question of evolution's validity. But if proven to be scientifically true, he noted that it would not contradict any of the church's teachings. By the late 20th century, another pope would accept evolution "as an effectively proven fact."[21]

Pope John Paul II would revisit the question of evolution in a 1996 message to the Pontifical Academy of Sciences. John Paul was a broadly read man compared to Pius XII, and he was seen as one who embraced science and reason. Regarding evolution, John Paul said, "It is indeed remarkable that this theory has been progressively accepted by researchers, following a series of discoveries in various fields of knowledge. The convergence, neither sought nor fabricated, of the results of work that was conducted independently is in itself a significant argument in favor of the theory."[22]

When Pope Francis addressed the Pontifical Academy of Sciences in 2014, it was not at all surprising to hear him say that the image of God the Creator as "a magician with a wand" was simpleminded. Science had much to tell us, and none of the things it tells us, he suggested, is inconsistent with the Catholic faith. "The Big Bang, which is today posited as the origin of the world, does not contradict the divine act of creation; rather, it requires it," he stated. Similarly, the pope argued that "evolution of nature is not inconsistent with the notion of creation because evolution pre-supposes the creation of beings which evolve."[23]

There is no doubt that many practicing Catholics still adhere to creationist views. In fact, a Pew survey in 2014 showed 26 percent of Catholics identified as creationists.[24] Most Catholics in 1925 probably sided with the prosecution in the Scopes Monkey Trial. But generations of practicing Catholics have grown up without the need to deny evolution or to deny science in the defense of their faith. The official Catholic doctrine accepts the science of evolution, although it sees God as the source or "designer" of the evolutionary process. This is a view that is compatible with that of the largest segment of Americans who accept evolution as a scientific fact. Remember, roughly only one in five Americans accept the science of evolution without the influence of the hand of God. About another 30 percent accept evolution with a presumed and necessary "assist" from God. The point is, as the Catholic example demonstrates, it was not a given that religion *had* to resist or deny science to protect or preserve itself. A majority of Americans (50 to 60 percent, depending on the survey source) do accept the science. But all surveys show that it is only a small minority of Americans who fully accept the scientific explanation alone (i.e., minus guidance by the Supreme Being) for the origins of human life. A much larger minority, at or approaching 40 percent in most surveys, subscribe to the creationist view.[25]

The scientific method often leads us to the discovery of counterintuitive truths. In fact, these truths can often be mind-blowingly difficult for us to accept. Galileo is a prime example of this phenomenon. It was not only the Catholic Church that had difficulty accepting his claim that the earth spins on its axis and orbits the sun. In other words, his teachings were contrary not only to church doctrine but to the common sense or thinking of his age as well. After all, one can't feel the earth spinning. To the average observer it sure looked like the sun went around the earth. Galileo's discoveries simply defied the common perceptions of his time. They were counterintuitive. But they were also true. Their being true did not make it easy for people to accept them, and scientific discovery is often reflexively rejected when it challenges what we have always believed.

It is often said that science is a method for deciding whether what we believe to be true actually has a foundation in the laws of nature or not. But the scientific method for deciding these things does not come to us naturally. In fact, it often conflicts with our tendency to cling to our intuitions or what some researchers call our native beliefs. Studies have shown that when the truth is counterintuitive, even when accepted as true, it is more difficult for us to process. One such study demonstrated that students with an advanced science education had a hitch in their mental gait when asked to agree or disagree that humans are descended from sea animals or that earth goes around the sun. Students who agreed with both statements, who answered correctly if you will, were measurably slower in answering than they were when responding correctly when asked whether humans are descended from tree-dwelling creatures or whether the moon goes around the earth. The reason for the slower response to the first two questions was that the answers are counterintuitive, and students, even knowing the correct answer, had to stop and think to process it. The second two questions (regarding humans descending from tree animals and the moon going around the earth) were easier for students to process quickly because they were both true and intuitive.[26] The major takeaway from this study was that even as we become scientifically literate, we still have to work to repress our naive beliefs. We never quite eliminate their influence entirely. They continue to lurk in our subconscious brains and slow us down, in spite of our acquired knowledge, as we try to make sense of the physical world.[27]

With respect to religion, the natural human tendency to cling to intuition or naive belief is no doubt enhanced. But that does not necessarily mean that our scientific thinking is always superior to our religious thinking. The evidence that supports scientific claims may be quantitatively and qualitatively superior to that for supernatural claims, but that may not always matter as much as we would like to think. One study found that students may not fully appreciate this qualitative difference. This might be because both types of claims (scientific and religious) are learned through testimony rather than firsthand observation.[28] In this particular study, students' scientific beliefs were compared with their supernatural beliefs. Student confidence in both types of beliefs were not associated with their ability to cite evidence in support of, or in conflict with, those beliefs. This study suggested that for many students scientific beliefs were qualitatively similar to their supernatural beliefs.[29] The ability to cite evidence in support of one's beliefs might be considered a form of scientific literacy. This study suggested that this ability is somewhat lacking even among individuals who have received multiple years of basic science education. The relationship between conceptual understanding and evidential reasoning is one that must be developed and that runs contrary to the way our untrained minds naturally or intuitively work.

Developmental psychologists tell us that there are two cognitive biases observable even in early childhood that can explain the intuitive appeal of something like intelligent design. The first is a belief that species are defined by an internal quality that cannot be changed. This is referred to as psychological essentialism. The second is that all things are designed for a purpose. This is a sort of a promiscuous teleology. The first bias says that members of a species have an inner inviolable "essence" that makes them what they are. This psychological essentialism is a powerful learning tool. In the face of all the visual evidence to the contrary, it allows us to easily group Chihuahuas and Great Danes into the same category (i.e., dogs). Teleology is the tendency to explain things in terms of their function rather than what has caused them. As applied to the physical world, even natural phenomena are defined in terms of their purposes. Ask a child why a rock is pointy, and the answer might be so that animals don't sit on it. The common-sense bias to believe that everything exists, or is the way it is, for a purpose underpins the intuitive attractiveness of intelligent design. At a minimum, it weakens our ability to accept the science of evolution without a godly assist.[30] Both biases (psychological essentialism and promiscuous teleology) interact with and are reinforced by cultural beliefs such as religion. But we must be careful to note that these biases are just as prevalent in children raised in secular societies. However they come to exist, these intuitive beliefs or reactions become increasingly entrenched over time. This makes formal scientific instruction more and more difficult as children get older.

The bottom line is that scientific thinking has to be taught. To be perfectly honest, it's often not taught very well. Students often do not fully comprehend science as a method; instead, they often come away thinking of science as a collection of facts. At the most basic level, many students don't really understand what evidence is. The scientific method is something we learn, not something we naturally comprehend. If you think about it, democracy is also something we had to learn. It was not a natural trait in our human development. For most of human history, neither the scientific method nor democracy existed. Human history is, even with its scientific and political advances, a fairly repetitive cycle that has seen us killing each other to get on a throne, praying to a rain god, or accumulating stuff for the sake of accumulation. For better and for worse, we really have done things pretty much as our primitive ancestors did even as we have modernized and perfected the methods for doing it. It can be argued that it is our science that has made us the dominant organisms on the planet, apologies to all other creatures great and small, and that our science has been the foundation for humanity's progress. It can also be argued that our imperfect mastery of science, the conflict between it and our material and often selfish interests, and our rejection of it in the name of values and beliefs we hold to be superior to it are among the things that pose the greatest danger to the future of humanity. In this context, and in light of the discussion we have had about the resistance to science, it is more than worth our time to discuss why this matters.

What Is the Cost of Resistance?

Let us begin with a basic question: How old is planet Earth? Is this question answerable, and, if so, on what basis may it be answered? There are, as you may already know, at least two answers to this question. Science says Earth is 4.54 billion years old, give or take a few years. Creationists, or many of them, say Earth is 6,000 years old. It might not have escaped notice that there is a bit of a discrepancy between these two answers. It might thus be important to understand how the creationists and the scientists arrived at their answers if we are to comprehend this discrepancy.

Creationists say Earth is about 6,000 years old. Their answer comes from the work of James Ussher, bishop in the Church of Ireland from 1625 to 1656. Archbishop Ussher took the genealogies of Genesis, starting with Adam in the Garden of Eden, and, assuming they were complete, calculated all the years to arrive at a date for the creation of Earth. According to his calculations, Earth was created on Sunday, October 23, 4004 BCE.[31] Creationists thus interpret the Genesis account to mean that the universe was created 6,000 years ago. Creationists believe that any evidence not supporting their theory is incorrectly applied, or that any data to the contrary is misinterpreted or manipulated. Their view is that the Bible is the only source that should be examined to prove the age of the planet, and the events recorded in it should be taken literally and as they interpret them.

Science says Earth is about 4.54 billion years old. Its answer is arrived at somewhat differently than the creationist account. The scientific process, simply stated, consists of finding the oldest piece of rock and measuring its age. The process of figuring out a rock's age often falls to the scientific techniques of radiometric dating, the most famous of which is radiocarbon dating. Scientists have used this approach to determine the time required for the isotopes in the earth's oldest lead ores, of which there are only a few, to evolve from their primordial composition, as measured in uranium-free phases of iron meteorites, to their composition at the time these lead ores separated from their mantle reservoirs. These calculations result in an age for Earth and meteorites and, hence, the solar system, of 4.54 billion years.[32]

Another way of looking at these two answers to the question of how old Earth is might look like this. Earth is about 4.5 billion years old. This is not a matter of opinion or belief; it is a scientific fact. This is something science can measure much like we can measure distance. The distance between Raleigh, North Carolina, and Milwaukee, Wisconsin, is 906 miles. Claiming Earth is just 6,000 years old is the mathematical equivalent of saying that the distance between Raleigh and Milwaukee is about six feet. If one were to accept a literal interpretation of the Book of Genesis and adopt what has been called the young earth form of creationism (i.e., the planet is 6,000 years old), one would have to willfully ignore or deny scientific knowledge on a very broad scale. It is not just evolution that bites the dust. Among the things that would be incompatible and in all probability denied by the strictest form of creationism would be scientific findings in a large number of fields that provide evidence that the planet is much older than 6,000 years. This would include findings in astrophysics, cosmology, nuclear physics, Newtonian mechanics, electromagnetism, fluid mechanics, heat transfer, mass transfer, reaction kinetics, thermodynamics, botany, genetics, immunology, neuroscience, trigonometry, cellular automata, geomorphology, plate tectonics, petrology, stratigraphy, volcanology, paleontology, metrology, and many more. This, if one thinks about it for just a bit, suggests that it requires a boatload of willful ignorance to sustain a strict creationist view.

Biblical creationists believe that humans and dinosaurs lived at the same time because the Bible says that God created both man and land animals on day six. The scientific consensus is that dinosaurs lived tens of millions of years ago and tens of millions of years before the first humans emerged. Humans have walked the earth for around 190,000 years. This is a mere blip in Earth's 4.5-billion-year history. Did people and dinosaurs live at the same time? No! After the dinosaurs died out, over 64 million years passed before people appeared on Earth. In other words, *The Flintstones* is not history. But the scientific evidence does not always win the debate with unscientific reasoning.[33] This is an important problem to be understood and analyzed if we care much about the future of humanity.

Some will take the increasingly popular view that science and religion are complementary ways of knowing about ourselves and the universe. As the Catholic Church has shown, one need not deny the science of evolution to retain one's faith. Science can be made to be compatible with faith. Others will say that this idea doesn't have a prayer of being true. Science, they will say, generates evidence-based knowledge, while religious faith consists of unverifiable supernatural convictions or superstitions. Some will say scientists are arrogant and foolish to be dismissive of religion. Others will say religious faith is delusional, and religious believers are dangerously intolerant toward nonbelievers and the inconvenient scientific findings that contradict their faith. This is precisely the sort of discussion we do not want to have, as we consider the true impact of the collision between evidence-based scientific truth and unscientific belief. It creates a situation ripe for deepening hostility and closed-minded argumentation in defense of one position or the other. It may be healthier to begin the conversation by understanding the difference between evidence and belief.

Science and religion are very different things. Science is based on the meticulous observation of nature. Scientists want to understand how the physical universe works. They assume that there are identifiable natural causes that explain things. This does not imply that some scientists are not, or cannot be, persons of faith. But those who believe that one or more deities exist typically assume that he/she/it/they do not interfere with nature. Neither is it implied that scientists have explained everything there is to know about the physical universe. In any given scientific area, a general consensus exists about most of what we might call the fundamental scientific truths. But there always remain unanswered questions as well. These unanswered questions may even lead to arguments or disagreements among scientists. These arguments and disagreements exist at the frontiers of each area of science. This is where new discoveries are being made, interpreted, and debated, and these debates can only be settled by uncovering, testing, and proving the reliability of evidence.

Scientists seek to understand the functioning of the physical universe through the use of the scientific method. The scientific method involves more than simply collecting data about a phenomenon. It postulates possible tentative opinions (hypotheses) about how a phenomenon may be explained or understood, tests each hypothesis to see if it is accurate, rejects it if it is not accurate, and recycles the entire process to replicate or test the reliability of results. Scientists publish the results of their work in peer-reviewed journals so that other scientists can evaluate their findings and test their validity. The scientific method is a process that, when properly and reliably executed, will result in ideas that have predictive power and that lead to greater understanding. After many cycles and general acceptance of the evidence-based conclusions within the scientific community, the conclusions that are reached (i.e., the most objective explanation of the physical universe that is possible) rise to the level of a theory, as in the theory of evolution. As we have previously stated, a scientific theory is essentially a proven fact.

Religious beliefs are very different from scientific evidence or facts; in that, they are typically based on faith. Most religious beliefs begin with the assumption that, through some form of revelation, God has taught the faithful an absolute truth. Any compromise of religious belief with the "beliefs" of scientists (i.e., the evidence-based reasoning of science) seems to the faithful to require them to reject their own religious beliefs. Few, of course, are willing to do that. There are many thousands of religious organizations in the United States. Of course, these groups are not unanimous in their faith. They hold diverse and often conflicting beliefs concerning deity, humanity, and the nature of the universe. Many consider their own faith to be the only completely true one. They believe that God revealed their unique and true faith to humanity in the form of the sacred and ancient books handed down over the generations. Many believe that religions other than their own (i.e., the "one true one") are man-made. They believe that the consensus of scientists on any matter is to be ignored if it contradicts their faith. The word of their God, as interpreted and understood by their religion and its traditions, is unerring.

Given the very different ways that science and religion define and arrive at "truth," they will inevitably disagree from time to time. Once upon a time, ancient Babylonian priests taught that lunar eclipses were caused by the restlessness of the gods. They were considered evil omens that were directed against their kings. Local astronomers eventually discovered that eclipses had a natural cause. This discovery did not change the superstitious beliefs of the priests, of course, but it did explain the phenomenon. In situations like this, history shows that science and religion often present us with conflicting interpretations of the same topic; this often places great strain on a culture and may cause needless suffering, retard much-needed progress, and even generate loss of life. This cultural strain, and its possible negative impacts, makes the resistance of science by religion or by any other entity a matter of broad public concern.

Resistance to science, whatever its source, leads to science denial. Science denial is often very costly and very dangerous. For example, vaccinations to prevent diseases are one of humankind's greatest achievements. Fear mongering and denying the science that demonstrates their safety may lead to undesirable and expensive outcomes. In 2011, a needless measles outbreak occurred due to the manufactured anxiety about vaccinations. Many parents refused to vaccinate their children. This caused unnecessary illness and more anxiety. It also cost the public money. Sixteen measles outbreaks involving a total of 107 cases were found by doctors at the Centers for Disease Control and Prevention to have created a total economic burden on local and state public health institutions of an estimated $2.7 million to $5.3 million.[34] As another example, a 2014 report sponsored by the nonprofit, bipartisan Risky Business Project sought to quantify some of the impacts of climate change on just the United States. It is found that if we continue on our current path, by 2050 between $66 billion and $106 billion of existing coastal property will likely be below sea level. By 2100 the numbers go to $238 billion to $507 billion. Crop losses in some parts of the country of up to 50 percent to 70 percent were projected.[35] Many more such costs and impacts were computed. Clearly, science denial can lead to problems. It may jeopardize our health and the future of our planet. It will also cost us a fortune.

Science has created high-paying and productive jobs in the United States. Many say science is the best, perhaps the only, way to create them in the future. Objectively speaking, science has a far better record of explaining both the world and how things actually work than does religion, ideology, or any value-laden belief system. Despite this, ever-greater percentages of Americans seem to be choosing religious and/or ideological rationales and explanations over science. Perhaps beliefs are more comfortable than facts, and we resort to them to avoid unpleasant or difficult things. Perhaps we find it emotionally difficult to accept facts that collide with the beliefs that make us comfortable. Of course, those who deny science certainly would not, one might presume, turn in their car for a horse and buggy. Nor would they prefer the medicine of 1850 to that of today, no matter how much they complain about the costs or mistrust doctors. They embrace all the physical advantages of science while rejecting the methodology . . . and anything else in science that conflicts with their religion or their beliefs.

Let us be clear that none of what is said in this chapter is meant to criticize religion or persons of faith. Neither is it meant to suggest that religion should be expected to operate only on the basis of scientific rules of evidence. It must simply be acknowledged that religion and science are very different things. This difference too often generates a conflict between fact and faith that is both unnecessary and quite harmful to humanity. It might be said that when faith and science collide, science ultimately wins. Why? Well, facts are inherently verifiable, whereas beliefs are not. But it is not this simple really. The process of reconciling paradigm-busting facts with long-cherished beliefs is agonizing and slow. With respect to evolution, we continue to struggle with the disagreement between faith and science almost 160 years after Darwin's publication and 100 years after the Monkey Trial. It continues to animate political discussion, generate creationist policy proposals, and influence voting behavior in the United States. We might also, while we are at it, remember that it was not until 350 years after Galileo's death that the Vatican officially apologized for the

Catholic Church's harsh treatment of the greatest scientist of his lifetime. Yes, the facts may ultimately win. But do they win before it is too late? Do they win before irreversible harms are done?

Ultimately the resistance to science for religious or ideological reasons leads to the outright denial of science. Science denial may be seen by some as a sign of ignorance, but for many citizens and politicians, it is both acceptable and applauded. Studies demonstrate that both ideology and religious faith contribute to or correlate with science denial in public opinion. Social scientific analysis has shown, for example, that political partisanship, party identifications, and ideology are the most important factors in explaining public opinion about climate change.[36] A 2010 study found that opinions about climate change were not influenced by scientific fact or the level of scientific education. Acceptance of the science about climate change increased with higher levels of education among Democrats, but it decreased with higher education levels among Republicans. In other words, the higher the education level of Democrats, the more they accepted the scientific conclusions about global warming, and the higher the education level of Republicans, the less they accepted it. These findings have been supported by other polls as well. This tells us that the divide in public opinion has less to do with science and more to do with emotions and values. Among American conservatives there is a great sense of mistrust and suspicion for what they perceive to be the liberal scientific elite.[37] But this is not the product of anything scientists are doing; it is the result of what conservatives fear about what the findings of science might imply in the context of public policy.

As is the case with political ideology, religion shapes public attitudes and opinions about science. A recent study examined issues involving science that have been contested in recent public debate, including human evolution, stem-cell research, and climate change. Not surprisingly, religious variables were found to be generally strong predictors of attitudes toward each of these individual issues,[38] even more so than political partisanship. This last finding may be subject to some qualification as there is a strong linkage between religiosity and conservatism, between religiosity and ideology, and between religiosity, conservatism, and Republican Party identification. In a 2015 Gallup survey that examined the relationship between ideology and religiosity among Republicans and Republican-leaning independents (see Table 5.2), the results showed how much Republicans in general skew both religious and conservative. Overall, 50 percent of Republicans were highly religious compared to the national average of 40 percent, and 61 percent were conservative compared to the national average of 35 percent. Not all highly religious

Table 5.2   Religiosity and Ideology among Republicans and Republican-Leaning Independents

Conservative/Highly Religious	34%
Conservative/Moderately Religious	15%
Conservative/Not Religious	10%
Moderate/Liberal/Highly Religious	14%
Moderate/Liberal/Moderately Religious	12%
Moderate/Liberal/Not Religious	12%

Source: Gallup, http://www.gallup.com/opinion/polling-matters/182210/highly
-religious-conservative-republicans.aspx

Republicans were conservative and vice versa. In fact, only a little more than one-third of Republicans could be classified in this survey as both conservative and highly religious. When one adds the number of Republicans classified as moderately religious, this increases the linkage between conservative ideology and religiosity enough to support the basic conclusion that Republicans skew both religious and conservative.

Republicans in general have a disproportionate percentage of conservative and highly religious Americans in their ranks, and highly religious people tend to be politically conservative. At a minimum, it may be said that a significant part of the Republican and conservative base is religious. Given that science denial is often associated with religion, and given that a large portion of the Republican base is highly religious, it should come as no surprise that most of the science denial by political candidates and elected officials is found among Republicans. In other words, Republican candidates and office holders have an electoral incentive to deny science. This means that religious attitudes about science and the issues that relate to it may have a potentially profound impact on public policy.

Religious attitudes are associated with public attitudes about all issues where science is an important component. Researchers have shown evidence that biblical literalists increasingly oppose the teachings of science where it introduces facts that conflict with their religious views. Speaking generally, Christian fundamentalists are less likely to have confidence in science and scientists than their mainstream brethren or the nonreligious. Conversely, respondents who believe that human evolution is "true" are more supportive of funding for scientific research than those who view evolution as "false" or "don't know."[39] With respect to climate change, there seems to be a correlation between religious fundamentalism and climate-change denial. It would appear, for example, that evolution deniers are also climate-change deniers.[40]

The impulse of highly religious or ideological individuals to deny what science tells them is understandable on a very basic level. We are all guilty of resisting facts from time to time. We are all more than capable of what is known as motivated reasoning. When we are presented with evidence that we agree with, that confirms our beliefs and our values, we tend to accept it uncritically. When presented with evidence that contradicts our beliefs and values, we often subject it to withering scrutiny, ignore it altogether, criticize the source, argue with the presenter, intimidate the presenter, and seek to generally discredit the evidence. This is an all too natural reaction that we fall prey to in the course of our daily lives. Our motivated reasoning is not scientific. It is not about finding the truth or creating knowledge. It is about winning an argument. While that may be a natural response, we also know that an educated response is very different in nature. Often, that knowledge enables us to catch ourselves, pause, and perhaps question our own motivated reasoning long enough to examine the new evidence in a different and more analytical light. This behavior is perhaps not the norm in most of our daily exchanges with the world, but it may be engaged just enough of the time to teach us a few new things and, as a bonus, help us maintain a reasonable semblance of sanity.

Motivated reasoning, while natural and understandable, is a roadblock to problem solving. In its worst form, it is absolutely toxic. As a habit of mind, it is certainly not the preferred mode of reasoning where important matters are to be decided and where the decisions have serious consequences. It may feed the resistance impulse and create conflict in times that require intelligent thought and collaboration. Climate-change deniers, for example, may call climate science believers "atheists" because they dare to believe that humans can "control" the climate. "That's God's job." From there the conversation only deteriorates. Clearly a different type of communication is needed, one that is the product of a very different habit of the mind, a habit that must be carefully nurtured and developed. It is not at all natural to us. While motivated reasoning is natural to us, it is most often not very good for us.

Motivated reasoning is actually the product of emotion (or what researchers often call "affect"). Our positive or negative feelings about people, things, and ideas arise much more rapidly than our conscious thoughts do. Reasoning comes later and works slower, but even it doesn't take place in an emotional vacuum. Our quick-fire emotions can set us on a course of thinking that's highly biased, especially on topics we care a great deal about, including those that touch upon our most personal and value-laden beliefs in areas such as religion and partisan politics. Even when we think we are reasoning, we may merely be rationalizing. In other words, our reasoning is motivated by our very emotional beliefs and values. Our "reasoning" is a means to a predetermined end that we've already uncritically accepted based on our beliefs and values. Basically, we are thinking and behaving like lawyers trying to win our case. Of course, our case is often shot through with subconscious biases, including confirmation bias, in which we give greater heed to evidence and arguments that bolster our beliefs, and disconfirmation bias, in which we expend disproportionate energy trying to debunk or refute views and arguments that we find unacceptable or disagreeable.

While motivated reasoning is not quite the same as what is meant by the newly coined term "alternative facts," it is a close relative. "Alternative facts" is a term that means lies or falsehoods that are used or promoted by people who don't like certain facts and intentionally proceed to replace them with made-up lies. The term was coined in the early days of the Donald Trump administration by Kellyanne Conway when she attempted to defend obviously false statements made by Sean Spicer, the White House press secretary. Motivated reasoning gives the appearance of addressing facts, even if its purpose is to disprove or dismiss them. Alternative facts simply ignore actual facts altogether. Instead of reasoning to a false value- or emotion-driven conclusion as motivated reasoning does, alternative facts simply assert something that is demonstrably not true on the face of it, but both do end up in the same place and reside in the same neighborhood. Their goal is the same. It is to deny, distort, distract us from, or reshape the facts. In a sense, "alternative facts" are the intellectually lazy person's substitute for motivated reasoning. In the end, however, both lead to a disaster where the public interest is concerned. They hinder genuine problem solving and lead to public and private decisions that transform challenges into full-fledged disasters.

Consider the person who has heard about a scientific discovery that confirms the theory of evolution and sees this as a challenge to his belief in divine creation. This person will have an immediate and subconscious negative reaction to this new information. This will trigger what happens next in the conscious mind. He will retrieve thoughts and "information" consistent with his beliefs and use them to construct an argument to challenge what he is hearing. The same holds true for, say, the climate-change denier. Upon hearing about a scientific study that confirms the scientific consensus regarding anthropogenic climate change, she will have an immediate and subconscious negative reaction that will trigger the retrieval of "information" consistent with her beliefs, and she will construct an argument to challenge what she is hearing. In this case, it is probably political ideology rather than religion that provokes the subconscious response, but it too is the product of values and emotions just as is the creationist's reaction to evolution. In fairness, it must also be noted that people who accept the theory of evolution or who do not deny climate change are capable of motivated reasoning as well. All of us are, and we do engage it more often than we would perhaps like to admit. The point is, science is often in competition with motivated reasoning of all types, and this too often leads to science denial.

Science denial may be seen as the product of motivated reasoning. Whether it is religion, partisan political ideology, or some other value that is at the root of our response, our resistance to scientific discoveries and our denial of what they say to us is ultimately an avoidance of a reality we do not like. We should strive to become aware of our subconscious triggers and to avoid becoming locked into predetermined conclusions where important matters of public policy are concerned. Motivated reasoning in politics and public policy making predictably impedes deliberations, negotiations, and all forms of collective decision making. It serves to increase the intensity and polarization of disagreement, and it magnifies the stakes that individuals and/or competing groups feel in defending their respective positions. Most seriously of all, the denial of science prevents us from using the knowledge it provides to best advantage. Denying the science of climate change, for example, means that policies related to energy, the environment, and the planet will be made absent the best understanding of what their true impact will be with respect to climate. This could very well increase the negative impacts of climate change and at great economic, environmental, and human cost. The waging of any policy debate on the basis of motivated reasoning makes for an intense competition, a polarization of partisans, and a public dialogue aimed at winning an argument as opposed to understanding and solving a problem. This is, of course, the norm in our public and political discourse in the United States. But science works differently and should be understood and treated differently.

It cannot be repeated often enough that science is a method. As we repeatedly emphasize, scientific reasoning includes the formation of hypotheses and their validation through the scientific method. The cycle of scientific reasoning begins with observation. Looking at a natural phenomenon, carefully observing it over time, leads to the formulation of scientific questions and hypotheses. This is where science provides possible explanations of a phenomenon and the laws of nature governing it. These possible alternative explanations, treated as predictive hypotheses, are tested. The testing is often gradual, taking place over an extended period. The evidence gathered may confirm or disprove the hypotheses.

This process is, most importantly, cyclical, meaning that as experimental results accept or refute hypotheses, these are applied to the real-world observations, and future scientists can build upon these observations to generate further theories.

Scientific reasoning is sometimes inductive and sometimes deductive. Inductive reasoning means that premises are based on facts or observations. Scientific reasoning may also be deductive, meaning that the conclusion logically follows from the premises and that the conclusion has to be true if the premises are true. Inductive reasoning is used when generating hypotheses, formulating theories, and discovering relationships, and it is essential for scientific discovery. Scientists use inductive reasoning to formulate hypotheses and theories and deductive reasoning when applying them to specific situations. Inductive reasoning is often called bottom-up reasoning and is used in applications that involve prediction or forecasting of phenomena or behavior. As an example, you may say that every tornado you have ever seen in the United States has rotated counterclockwise. You see a tornado off in the distance, and you are in the United States. You predict, based on previous repeated observation and study, that the tornado you are now seeing is rotating counterclockwise. The statistical probability of this prediction being correct is in fact very, very high. Most tornadoes in the United States rotate counterclockwise, but not all of them do. Therefore, the conclusion is probably true but not necessarily true. Inductive reasoning is, unlike deductive reasoning, not logically rigorous. Imperfection can exist, and inaccurate conclusions can, but very rarely do, occur. Inductive conclusions are right most of the time. Deductive reasoning is sometimes referred to as top-down logic. The classic example from Aristotle, the father of deductive reason, is his proof that Socrates is mortal: "All men are mortal; Socrates is a man. Therefore, Socrates is mortal." The premises of the argument (all men are mortal; Socrates is a man) are self-evidently true. Because the premises establish that Socrates is a member of a group of mortals, it is obvious that Socrates is also mortal. In deductive reasoning, the conclusions are mathematically certain.

Scientific reasoning and motivated reasoning are very different things. It actually matters which type of reasoning is being employed in the assessment of any phenomenon. Scientific reasoning is evidence driven, and its conclusions are accurate. Motivated reasoning is emotion and value driven, and its conclusions are predetermined and not necessarily, very rarely in fact, supported by objective evidence. This is not to say that all scientific conclusions are 100 percent certain. Neither is it to say that there are no scientific debates or disagreements. But these debates, or disagreements, are subject to rigorous scientific testing, and their resolution must always be evidence driven. Indeed, when scientists arrive at a consensus regarding a natural phenomenon, we can be assured that it is the best factual explanation that is humanly possible. It is the truth. With respect to motivated reasoning, which quite frankly is the sort of reasoning typical in all cases of science denial and most cases of partisan argumentation in our politics, the conclusions reached are not very trustworthy at all. Simply stated, we are talking about facts and beliefs. This is simplistic, but it is straightforward. The facts in science remain unaffected by our beliefs. Our beliefs, on the other hand, may affect the way we react to the facts of science. If the facts cannot survive our beliefs to the contrary, science will be of no benefit to us whatsoever.

We tend to think of science as merely a tool. We separate it from all the value-laden questions we are presented with in our private and public lives. Science has developed, for example, the incredible surveillance technology that so easily and completely invades privacy and the personal freedom bestowed by the U.S. Constitution. The question of how that technology gets used is not considered a scientific one. People debate the justification of enhancing national security at the expense of personal freedom, not the science and technology behind it. But we must also think of science as a method and as a way of thinking that must be learned and developed because it is very different from our instinctual and motivated thinking. There are a growing number of issues and concerns, like climate change for example, where we cannot begin to define the problem or the choices to be made without scientific knowledge. Indeed, to comprehend the options and to develop the criteria for choosing between them requires application of the scientific method. But if we equate motivated reasoning with scientific reasoning in the context of our public debates about these issues where science is the necessary mode of thought to comprehend and respond to problems, how can we possibly hope to make the best decisions?

Conclusion

In and of itself, the evolutionism-creationism debate may not seem significant. But if one stops for a moment to reflect on this debate as a contest between motivated reasoning and scientific reasoning, it may begin to seem more significant. If we find it acceptable to suggest to schoolchildren that evolution is a "controversy" or that alternative "theories" such as creationism and intelligent design should be taught alongside evolution in our science curriculum, we are no longer teaching science or the scientific method; we are elevating motivated reasoning, of the religious or any other variety, to a level of equality with scientific reasoning. If we do that with regard to this debate, why not do it with all? Why maintain the status of and respect for science and the scientific method if we see no qualitative difference between scientific fact and religious belief in the study of the physical universe? If science can be questioned, doubted, rejected, resisted, or made to be the equal of belief, why should we even teach the scientific method? Bring on the alternative facts.

Why is there still, in the 21st century, such concern in so many schools about teaching evolution? There is a complete consensus among scientists all over the United States and the rest of the world that evolution is the backbone of modern biology. It is a demonstrable reality historically as well. Most people really don't understand science. Even those on your school faculty or state board of education often need much more education themselves before they can be trusted to govern how or what our kids will be taught. They don't know what science is, how it works, what hypotheses and theories are, or even the purpose behind it.

What should our reaction be to the question "Should we teach creationism or intelligent design in our public schools?" The first thought that should come to mind is, "Why in the world would we want to do this?" The United States is already falling behind on the list of industrial nations when it comes to science literacy. It is near the bottom of the list of industrialized nations when it comes to teaching evolution in our public schools. As a result, over 40 percent of American adults reject evolution and embrace creationism. While this may not seem so important, reflect on what this really means. It means the rejection of sound scientific reasoning is seen or accepted as legitimate. It sets the pattern of rejecting fact and supporting in its place a fairy tale. How can a nation hope to compete in a global market if it is routinely willing to allow ancient or primitive fables to compete with proven scientific facts in the assessment and understanding of the physical world?

One might argue that because the creationists and the intelligent designers keep losing in court that we will come out all right in the end, but the creationists have had much more success in the cultural arena. They still infiltrate the public schools, where it is not unknown for creationist teachers to advance their religious views in class. Creationist churches instruct children to refuse to learn about any science their religion rejects. While some of us may regard this as child abuse, others may be willing to say to each their own. Consider the student who receives a test question in her biology class that asks her to explain two forms of evidence supporting evolutionary change and natural selection. Is it a good thing for her to feel empowered and correct to respond by saying, "I refuse to answer because I don't believe in evolution"? Do we want her to grow up thinking that there is no difference between science and belief or that belief may disprove science or be a valid reason to resist and reject it?

The resistance to science that is all too often born of religious or cultural values is, if one thinks about it at a very basic level, not at all rational. Science can neither prove nor disprove the existence of God. It has neither the desire nor the capacity to engage that subject. What it seeks is to arrive at the best evidence-driven understanding of the physical universe, and it does a very good job of doing that. Religious and cultural values, and the motivated reasoning that follows from them, cannot prove or disprove the laws of physics or render any legitimate assessment of the physical universe that competes with science. To use religious belief to deny science is simply not a reality-based foundation for reasoning. Likewise, using science to pass judgment on the validity of anyone's religious faith is not appropriate either. Science might be able to say that scientific analysis does not support some of the specific things that people of faith believe, but it cannot prove that what they generally believe about a Supreme Being is without validity. It can only say that the belief has no proven scientific validity. But the number of things that are inexplicable by science leaves plenty of room for faith. Likewise, the number of things that are explicable by science leaves no room for one to dismiss facts for fantasy.

Scientific reasoning and faith are, as we have attempted to explain, two distinct and unrelated things. They represent two very distinct and incompatible types of reasoning. As such, it is never truly appropriate to use one to bash or deny the other. That said, it may nevertheless be possible to suggest that what we believe, whether provable or not, should not deny the reality of the things and processes that are in fact knowable and provable. That would simply not be a sane thing to do. In this light, it can be observed that most of the world's religions are of primitive origins. Their holy books and the precepts of their faith are the product of primitive minds. They were written and established by primitive people for primitive communities. They predate most of the scientific discoveries that make the physical universe knowable and understandable for us today. A strict adherence to ancient texts and primitive teachings is illogical from the perspective of science. But, just for fun, consider that it might also be illogical, perhaps sinful, from the perspective of a particular variety of "motivated" religious reasoning.

Take, as an example for discussion purposes, the concept of intelligent design. Science does not entirely deny the notion of intelligent design. It simply cannot prove it. Thus, it has no opinion on the matter. What science can say is that the big bang is where it all began. The big bang theory posits that the universe as we know it started 13.8 billion years ago with a small singularity and then continued inflating over billions of years to the cosmos that we know today. Our current instruments do not allow astronomers to peer all the way back at the universe's birth. Much of what we understand about the big bang theory comes from mathematical theory and models, including Albert Einstein's general theory of relativity along with standard theories of fundamental particles. NASA spacecraft such as the Hubble space telescope and the Spitzer space telescope are able to continue measuring the expansion of the universe as we speak. One of the goals has long been to decide whether the universe will continue to expand forever, or whether it will someday stop, turn around, and collapse in a "big crunch."

Bear in mind that the big bang is a scientific theory and not merely a hypothesis. This means the big bang theory has met all challenges and answered them and that all the known evidence explains the phenomenon in question. We know this. But *we do not know what caused the big bang*; science can only speculate about the cause. It will very likely never find concrete evidence for what caused it because this is outside the realm of our observation. This actually leaves room for intelligent design to be made whole with science. Think about that for just a moment. It enables the person of faith to reconcile what he or she believes about a Supreme Being or intelligent designer with science. The big bang for them becomes the act of their God or the designer. Science can neither prove nor disprove that belief and really has no interest in doing either. The person of faith, on the other hand, may accept modern science and interpret the creation myth of its holy book as a story not to be taken literally but figuratively. This is a rationalizing of one's faith, or a form of motivated reasoning, that arrives at a predetermined end (e.g., belief in a Supreme Being) but *does so by reconciling the belief with what is knowable and real as opposed to denying it*. There will be minimal and most likely no resistance to science in this case.

One could carry the motivated reasoning of a believer who is reconciled with science to some interesting levels. It actually could seem to be inconsistent with the concept of intelligent design, or the variation of it we have just discussed, to deny science or to adhere to ancient and primitive reasoning over the advances made by science. How so? Well, if one argues intelligent design, one might well explain the human ability to investigate the physical universe and to learn about it as a part of the design. One might argue that the investigator, the human scientist let us say, has been designed to explore, comprehend, and explain the cosmic design. While it may not be able to prove intelligent design, science can reveal the truth about the universe. Intelligent design might be said to assume a "they" for whom the truth is intelligible. The rendering of this intelligibility is not a given but is actually dependent on the cultivation of the habits of mind and objective methods of assessment that may, under the right circumstances and after much rigorous testing, produce it. The design is not fully revealed. It may not be scientifically discoverable. But the truth about the universe is discovered or revealed. Human scientific agency and the truth it is able to discover is part of the "design" and is the product of unique (God-given for the believer) habits of mind. The cultivation of these habits and methods might be considered the prerequisite for the faithful, for it is only in doing as the design requires (i.e., science) that the truth may be revealed. And the truth will set you free! To deny the science, the truth, to refuse to develop the habits of mind and the methods required for attaining the truth, would be to refuse to participate in the design. To refuse to participate is to choose primitive understanding over knowledge. This choice might be the definition of "sin" in this construction, for it ignores the design and refuses to use the tools the designer gave humanity to learn and to understand the truth. Well . . . it's fun to play with all this, of course, but the point is a serious one. Faith and science are different, but they needn't be seen as innately hostile to each other. In fact, that is the worst thing they could be. Finding a way for individuals to rationalize an agreement between the two, however motivated the reasoning, is the sanest alternative. The traditional young earth creationist variation on the theme represents a very different and irrational type of motivated reasoning. There is no way to reconcile it with science.

The creationist may well argue that "science" of some distorted and unscientific sort supports the view of a historical six-day creation, as outlined in the first chapters of Genesis. Some creationists even try to list a number of "scientists" who believe in the creationist model. There is, of course, no scientific basis for such views. The fact is that creationism, whether it be old-fashioned young earth creationism or its more progressive intelligent design descendant, is simply not science. There is no scientific debate about the correctness of evolution, and the scientific evidence disproves the young earth argument entirely. There can be no scientific debate about intelligent design unless or until the intelligent design advocates produce some science or scientists have the proof to refute it. Likewise, because there is not a scientific basis for proving or disproving intelligent design, science can neither be expected to accept nor to reject that which is not proven or provable. In fact, science cannot take any position on any matter without evidence. But young earth creationism is a matter that science can decidedly reach conclusions about on the basis of solid, irrefutable scientific evidence. There is no scientific debate here. In any rational debate between science and religion about evolution, science would win everywhere, except perhaps in the public imagination. It is a sad commentary that in the most scientifically advanced nation the world has ever known, creationists can still persuade politicians, judges, and ordinary citizens that evolution is a flawed, poorly supported fantasy. They lobby for creationism and/or intelligent design to be taught as alternatives to evolution in science classrooms. In science classrooms! This is to misunderstand what science is. To the extent that such efforts continue to be popular across the nation, they work against scientific literacy when it is arguably more important than ever in address-ing our national needs. Faith is a personal thing, and a basic right to believe is to be cherished and protected. But promoting science denial, resisting reality to defend faith, is an attack on reason that a sane society should really try to avoid. It should certainly never drive its politics.

As a matter of public policy, and as a matter of our national interest, our political and policy-making decisions should not be based on mere beliefs and opinionated values, religious or any other. Of course, that ideal is probably unachievable in a democracy. Opinions, values, and emotions are always at play and always influencing debates and policy outcomes. There is no escaping that, nor can there be really. However, with respect to science and politics, one might suggest the goal of any modern, advanced, intelligent society should be to never base its policy decisions on the denial of science. Conscious efforts to resist and deny science in order to adhere to what one believes and to dismiss inconve-nient facts is illogical at least. It is also undesirable and potentially harm-ful and even deadly. But this resistance and denial will inevitably happen in our politics and impact it in a free society. That said, it is an altogether different matter when those who are elected to serve enable this resis-tance and denial. Worse still is the silence or acceptance of this by all, including scientists, who actually do know better. Scientists, public offi-cials, educators, and informed citizens all bear some of the responsibility for what can only be called an alarmingly persistent tendency to resist science and to prefer belief over fact. This is the 21st century. This is almost 100 years after the Monkey Trial. Yet, even in this remarkable and advanced age, there are simply too many indications that Americans remain a primitive people with all the latest inventions.

CHAPTER SIX

Panic Reflex: Pandemics, Science, and Politics

It infected an estimated 500 million people worldwide, about a third of the world's population at that time, and killed 30 to 50 million victims. Twenty-five percent of the American population became sick, and 675,000 died. It killed 10 times more Americans than did World War I.[1] One dramatic account of it reads as follows: "Those afflicted were first aware of a shivery twinge at breakfast. By lunchtime, their skin had turned a vivid purple, the colour of amethyst or the sinisterly beautiful shade of the heliotrope flower. By the evening, before there was time to lay the table for supper, death would have occurred, often caused by choking on the thick scarlet jelly that suddenly clogged the lungs."[2]

"It" was the Spanish flu pandemic of 1918. This influenza pandemic was first observed in Europe, the United States, and parts of Asia before swiftly spreading around the world. By August 1918, it had mutated, and an epidemic of unprecedented virulence exploded within a one-week period, affecting port cities around the globe.[3] Surprisingly, many flu victims were young, otherwise healthy adults. At the time, there were no effective drugs or vaccines to treat this killer flu strain or to prevent its spread, and it killed more people in less time than any disease before or since. In the United States, citizens were ordered to wear masks, and schools, theaters, and many public places were shuttered altogether. The Spanish influenza moved swiftly across the United States following the railroads, and it propagated fastest in the communities closest to them.[4]

As the Spanish flu spread across the United States, public officials were cautious, skeptical, and poorly informed. It goes without saying they were totally unprepared. The *Journal of the American Medical Association* opined at the time that the Spanish flu "should not cause great importance to be attached to it, nor arouse any greater fear than would influenza without the new name."[5] Unfortunately, public health officials and physicians had to undergo heavy doses of pandemic exposure before they understood the nature and the impact of the event. If public health officials and the medical community were slow to react, the governmental establishment was slower still. As the seriousness of the pandemic became apparent, the initial decision of those in positions of political power was to say as little as possible about it so as to prevent a public panic.[6] To an alarming extent, the American public was generally uninformed.

One would think that since 1918, given its many painful lessons, the United States would have developed the capacity to anticipate and prepare adequately for new and novel diseases. Unfortunately, while much progress has been made, there is still a ways to go. Most recent new diseases—hantavirus and West Nile fever are two that come to mind—were well established months or even decades before they appeared in numbers large enough to be identified.[7] A significant part of the problem is that public health surveillance of the human and animal populations has not improved sufficiently to meet all the challenges presented by new disease outbreaks. As a result, we remain highly vulnerable to the inadvertent exposure to higher risk for serious new diseases.[8] A pandemic on the order of the 1918 Spanish flu could actually be many times more serious today. We were much more self-sufficient in 1918. In today's corporate and free-trade environment, just-in-time inventory management and global supply chains are the norm. Given this, economic analysts predict that there would be a full-blown global economic collapse in the case of a worldwide pandemic. A complete shutdown of the global supply chain and unprecedented human suffering on a global scale would result.[9] Not a pretty picture, to say the least.

Is another flu pandemic likely? Yes. The experts tell us that the world is becoming increasingly likely to see a major pandemic. The highly predictable process works something like this: It starts with migration of agriculture and urban environments into more rural and remote areas. This increases the likelihood that a potential pandemic strain of a pathogen will come into contact with humans. Thanks to the increased densification of both animal and human populations, these pathogens can spread in a localized environment and evolve to cause greater problems. Finally, given the ease and frequency of modern travel, the pathogen can move from the localized area to practically anywhere in the world. Global travel increases the seriousness of the risk more than any other single factor. In

1918, it took months for the Spanish flu to circumnavigate the globe; today it would take a single day. Worldwide, experts agree we should expect a new pandemic. Many say we are overdue for one.[10] The question is, would we be prepared for one if it happened?

The world is, one might reasonably surmise, doing its best to be better prepared for a major outbreak. In the aftermath of the SARS outbreak in 2003, the H5N1 bird flu threat in 2005, and the H1N1 swine flu of 2009, there has been some significant global action. By 2011, 158 countries had produced improved pandemic-preparedness plans. In the United States, more money has been poured into the development of new vaccines and antiviral drugs. Researchers have a much better overall understanding of influenza and other risky pathogens and are able to identify viruses quicker than ever. New techniques allow them to explore troubling strains, and new and improved models are able to predict where a new disease might emerge and how it might spread. There have indeed been many advances and improvements, *but all this may not be nearly enough.* For example, we have not yet really managed to predict a new influenza outbreak. And each new outbreak provides a snapshot of our preparedness efforts. The unexpected H1N1 outbreak in 2009 exposed many problems that have caused ongoing concern, including the slow deployment of vaccines and simple breakdowns in communication. Thankfully, that virus was not especially deadly. Recent outbreaks of Ebola and Middle East respiratory syndrome have been instructive reminders that epidemics are frequent and often a surprise. They also highlight how, in spite of our renewed efforts, we remain generally more ill prepared to handle them than we'd like to admit. Scientists and political actors both feel the pressing need to answer one basic question: How do we improve our preparedness for public health emergencies?

National planning and preparedness for a serious pandemic threat or biological attack is a subject of great discussion in the United States and around the world. The level of readiness continues to be a matter of great concern. There were a number of essential national strategies, plans, and policies issued in the first decade of the 21st century to address the most serious outbreaks or evolving public health concerns. The two most notable and foundational documents in the United States may be the "National Strategy for Pandemic Influenza" of 2005 and the "National Strategy for Pandemic Influenza—Implementation Plan" of 2006. These documents have a broad focus and wide inclusion of international, federal, state, tribal, local, and private-sector partners for a threat that is likely to have the greatest global impact. Absent adequate preparedness, the panic and confusion that may result as a pandemic unfolds could lead to a health crisis that spirals out of control. We have not, despite the good efforts that have begun, done nearly enough to prepare for potential pandemics. There are gaps in our scientific defenses to be sure. But the number-one problem is that policy makers at all levels have not given these threats anything close to the priority consideration they demand.

Pandemic Preparedness

The fact that good plans have been developed does not automatically translate into successful and maintained preparedness. Often, when plans are developed, they are filed and forgotten as a new day brings new issues and concerns to the fore. Following a period of sustained pandemic planning that began in 2005, the Department of Homeland Security (DHS) did a follow-up study of pandemic preparedness in 2014. The DHS Office of Inspector General (OIG) issued a report entitled "DHS Has Not Effectively Managed Pandemic Personal Protective Equipment and Antiviral Medical Countermeasures."[11] The following is an excerpt from that report:

> DHS did not adequately conduct a needs assessment prior to purchasing pandemic preparedness supplies and then did not effectively manage its stockpile of pandemic personal protective equipment and antiviral medical countermeasures. Specifically, it did not have clear and documented methodologies to determine the types and quantities of personal protective equipment and antiviral medical countermeasures it purchased for workforce protection. The Department also did not develop and implement stockpile replenishment plans, sufficient inventory controls to monitor stockpiles, adequate contract oversight processes, or ensure compliance with Department guidelines. As a result, the Department has no assurance it has sufficient personal protective equipment and antiviral medical countermeasures for a pandemic response. In addition, we identified concerns related to the oversight of antibiotic medical countermeasures.[12]

The good news is that this OIG report led to upgrades and improvements in U.S. pandemic planning. This should remind us of the old adage that plans are nothing, but planning is everything. Constant review and revision is necessary, and, unfortunately, it is often slighted. As it turns out, the 2005–2006 pandemic planning process was a good one, but it was far from perfect. A closer look at it may help us to see both the importance of such planning and some of the inevitable shortcomings that may be avoided if we begin to improve the relationship between science and politics in the planning process.

Pandemics are viewed as health problems. Unlike security risks, which easily command the constant and vigilant attention of policy makers, political partisans, and citizens generally, health problems are relegated to public health departments or agencies where, absent an immediate threat, they are quickly forgotten by policy makers and totally ignored by the general public. Rather than having any urgency about building up defenses, as one would for a war or a terrorist threat, policy makers tend to give low priority to potential pandemics. But scientists, in the course of their ongoing work, bring major new threats to the surface from time to time. Actually, they monitor and bring new threats to the surface on an ongoing basis. New flu strains and threats emerge routinely. From Ebola in West Africa to Zika in South America, we have seen that dangerous disease outbreaks are on the rise worldwide. Since 1980, the number of new disease outbreaks has tripled. The U.S. Centers for Disease Control and Prevention (CDC) is constantly on the lookout for the flu strains that have the greatest potential to cause a pandemic. Although health experts and scientific researchers have improved our capacity to identify serious threats, nobody knows for sure which new flu strain or which mutation in an existing strain will cause the next pandemic. But the consensus is that there will be a next pandemic. The H5N1 outbreak was significant; in that, it stimulated a new sense of urgency about pandemic preparedness.

In 2005, world health experts and researchers who had been monitoring the potential for a new influenza pandemic began warning about a new and severe influenza virus. This virus was the H5N1 strain, better known as the bird flu or avian influenza. Between 2003 and 2008, this virus caused the largest worldwide poultry outbreak ever recorded. It is important to note that all flu strains originate in the animal population. Some spread from animals to humans and, eventually, may be transmitted from human to human. By all indications, experts say that the bird flu holds the potential to mutate into a very deadly threat to humans once a fully contagious virus emerges and becomes transmissible from human to human.[13] Other strains and threats have been identified since, including the H7N9 bird flu in 2017, but the H5N1 strain, still a concern over a decade later, was deemed concerning enough to stimulate a new and serious round of pandemic preparedness worldwide.

Between 2003 and 2008, 65 countries had experienced animal outbreaks of the H5N1 virus. This included 27 countries that were added to the list in 2007 alone. From 2003 to 2008, 14 countries had experienced confirmed human cases of H5N1, including 382 individual cases with 241 deaths for a death rate of 63 percent. Human-to-human transmission was thought to have already occurred in Indonesia, which had experienced the largest number of cases (133) and the highest death rate (81 percent).[14] The spread of the H5N1 virus in poultry and the spillover infections in humans ignited serious concerns about a possible new pandemic. Most scientific experts believe that a major pandemic, the most serious one in over a hundred years, is on the horizon and that the probability is very high over the next 20 or so years. It could very well result in anywhere between 5 million and 500 million deaths worldwide. The World Health Organization more conservatively estimates 1.9 million to 7.5 million deaths.[15] Fatality rates globally will vary according to the resources that national health organizations have in place. The potential for the H5N1 virus to mutate and become a full-fledged pandemic disease that would spread rapidly, affecting entire nations at once, was perceived as a very real threat. As noted, the probability of an avian flu pandemic is generally considered to be very high. While it may not be the H5N1 strain, many have concluded that an avian influenza will be the source of the next pandemic.[16]

As the threat of an influenza pandemic hit home in the middle of the first decade of the 21st century, it stirred governmental concern and action at all levels in the United States. The DHS developed a national strategy to guide state and local governments, as well as the private sector and individual citizens, in preparing for the most serious flu pandemic in almost 100 years. This national strategy called for the implementation of over 300 actions by federal departments and agencies. It communicated expectations for nonfederal entities (state and local governments, the private sector, critical infrastructure, individual citizens, etc.). The key responsibilities of the national government in pandemic preparedness include coordinating the response of federal agencies, funding and supporting the development of vaccines and antivirals, stockpiling vaccines and other countermeasures, and coordinating with other nations. State and local governments, as well as private businesses, also have specific planning responsibilities for pandemic preparedness and response. These too are specified in the DHS national strategy.

As the national pandemic planning process unfolded, most Americans were no doubt blissfully unaware of it. As we shall see later, there were also political twists and turns that impacted the amount and the quality of preparedness planning over the remainder of the decade. But, before discussing any of that, it may be instructive to examine in a bit more detail the H5N1-inspired pandemic planning process as it played out at the state and local level. This will provide us with a feel for some of the conversation that took place and some common challenges that the pandemic-preparedness efforts encountered.

State and local communities, as directed by the DHS, began planning for a possible avian influenza pandemic in 2005. These preparations revealed wide differences and little consensus about the best policies and strategies.[17] Among business- and private-sector actors, the effectiveness of planning and the quality of relationships necessary for the creation and implementation of pandemic plans remained questionable.[18] During this period, a majority of global corporations prepared and put in place avian flu pandemic plans. Small businesses, however, represented a special concern. A majority of small businesses had not done much, if any, planning,[19] in spite of the fact that they are among the organizations that would be the most vulnerable in the event of a pandemic outbreak. Whether they had planned or not, a major concern for businesses of all sizes was the apparent lack of coordination with the public sector. In an assessment of pandemic planning in 2006, 90 percent reported that they had not had discussions with government at any level.[20] Working to improve coordination between the public and private sectors would become a priority for pandemic planning to be effective.

The planning process revealed that agricultural communities had animal husbandry practices (i.e., factory farming) that must be considered in relation to pandemic threats. For example, the poultry industry's practice of keeping large numbers of foul in close proximity increases the opportunity for the transmission of disease. In general, the monitoring of bird and animal populations for signs of the H5N1 virus was highlighted as a critical public health surveillance function that needed to be improved. This applied all the more to meat-packing and poultry storage facilities. Better monitoring and coordination with these entities was stressed as essential.[21] The best chance any community has to control a pandemic is directly related to how quick it is to recognize that it is happening. To whatever extent is possible, the collaboration of public health, the medical community, agriculture, and business must be enhanced to promote improved surveillance of the human and animal populations for signs of disease. Ideally, pandemic planning must include some sort of early warning system.

Early warning is considered critical with respect to containing or delaying the spread of a pandemic disease, but local communities certainly cannot be solely responsible for early warnings. They cannot do nearly enough on their own initiative in this regard. Animal disease surveillance, for example, is a state responsibility. During the planning phase for the H5N1 virus, animal-health professionals cited the need to improve surveillance for animal diseases. Most states have laws requiring reporting or monitoring animal diseases, and these now include the possibility of H5N1 outbreaks in the animal population. Existing health information systems were targeted for improvement, and states are expected to provide more assistance and leadership for local communities.[22]

Public health surveillance of the human population is also necessary to provide early warning and information to decision makers. This requires the systematic collection and interpretation of data. A global pandemic would require an integrated worldwide network that brings health practitioners, researchers, and governments together across national boundaries. Local communities cannot be responsible for that, of course. This is why the national and state units worked, during the planning phase, to expand their capacity to provide that leadership and to do more to bring local communities into an integrated disease surveillance and response strategy. Local communities, it was thought, would benefit from incorporating into their planning an awareness of and participation in federal and state surveillance of animal and human populations. Likewise, entities such as the CDC and the WHO were considered valuable sources of information for communities as they planned at their level. The bottom line is, public health surveillance is one of the best weapons to be enhanced in the effort to avert epidemics and pandemics. The improvement of health-surveillance capacity was one of the major objectives of the H5N1 planning process.[23]

Vaccination is, of course, the most reliable way to limit the impact of a possible H5N1 pandemic, but that is beyond the control of local planners. It is the responsibility of the federal government and the drug industry to collaborate on a vaccination plan and implement it as quickly as possible. Local governments would, however, be critically involved in the distribution process. Local plans, or most of them, included the designation of public treatment centers and points of distribution for vaccines. But such a pandemic would likely emerge and engulf communities before an effective vaccine could be produced. In the short term, as a vaccine is developed, people would need to rely on available influenza medications, such as Tamiflu. It was found, as the planning process began, that the United States did not have nearly enough Tamiflu stockpiled to meet the anticipated demands of a pandemic scenario. Serious efforts have been made since to improve that situation. But even in a best-case scenario, shortages of antiviral medications must be anticipated.[24] Several H5N1 vaccines have been developed, approved, and stockpiled by a number of countries, including the United States, but the mutations that may lead to the H5N1 becoming a pandemic have not yet occurred, so there is no way of knowing if the vaccines that have been developed will be effective. They likely will not, and a new vaccine will have to be created.

In the H5N1 planning over the 2005–2006 period, most local plans followed the advice of the WHO that available medications and what was expected to be a limited supply of the new vaccine be first given to essential personnel as they became available.[25] Most plans included the following as essential personnel: medical personnel, public health and emergency responders, police, firefighters, and critical infrastructure personnel. Many also suggested or included workers in the transportation industry responsible for the delivery of foods and medicines. Other nations developed similar lists of essential personnel. Ominously perhaps, but logically if one assumes the worst, Australia placed funeral directors among those first in line.[26] The question of essential personnel aside, most of the local plans operated on the assumption that the general American public would be left without much protection at the onset of a serious pandemic. This led to the consideration of what might be called defensive strategies.

In the expected absence of sufficient vaccine and shortages of other medications, and with the probable overcrowding of hospitals in a pandemic crisis, local planners were expected to be prepared to employ defensive strategies as the pandemic threat level rises, including social distancing, respiratory etiquette, and other hygiene measures. School closures, public-gathering bans, and travel bans might well be required to slow the course of a pandemic, and the plans needed to identify the trigger mechanisms that would activate these measures.[27] But, truth be told, such measures may be of limited utility in a present-day pandemic scenario. It is probable that a community would be in the midst of a pandemic before these measures could be implemented. Given this, and the expected shortages of medical interventions and vaccines, most individuals and families would be expected to arrange for their own safety. This means that risk communication and public education would have to be an important part of the planning in communities across the country.

Communicating the risks associated with an H5N1 pandemic and educating the public about the appropriate and necessary self-defense measures to mitigate these risks was regarded as an essential component in all local pandemic planning. In 2007, the American Public Health Association conducted a survey to determine the extent to which Americans were prepared for a health emergency. The results demonstrated that too many Americans are unprepared. Only 14 percent had the three-day supply of food, water, and emergency supplies recommended by the American Red Cross; 32 percent had taken no action whatsoever to prepare for a public health emergency; and 87 percent were found to be generally unprepared or underprepared.[28] Another general finding indicated that roughly 4 in 10 employers did not believe that a public health emergency would ever impact their businesses. Regional food-distribution centers reported that they had spent considerable time and resources on preparing for a pandemic, but local pantries and food shelves indicated they had not begun to prepare.[29] Surveys such as this suggest that the task of educating the public to prepare and to take defensive measures for a possible avian influenza pandemic could be very daunting indeed.

Social scientists have been studying and analyzing people's responses to risk communication for decades. The possibility for a sound pandemic risk communication and public education strategy exists, to be sure. Unfortunately, it has been largely undeveloped.[30] While most plans addressed the need, the results were mixed. Ironically, very little effort was made during the planning process itself to include or inform the general public, but it was agreed by all that the effort to inform and educate the public was a critical necessity. The communication of risk and public education is complex and difficult and requires coordination among various types of experts, including subject-matter specialists (e.g., public health, social services, law, emergency management planners, and education), risk and decision analysts to identify information critical to the decisions of various audiences, psychologists to design messages and evaluate their success, and communications systems specialists to ensure that tested messages get communicated within the emergency response system.[31] It is probably safe to assume that no foolproof system for doing any of this has ever been designed. The recipes for failure are many, and the imperfections of even the best risk communication and public education plans are such that communication failures or breakdowns are a common feature of most emergency response activities.

A final component mentioned, although not really addressed in most local plans, was the ethical dimension of a pandemic threat. Planners recognized that the allocation of scarce medical supplies or vaccines, for example, or the application of control measures during a pandemic occurrence were ethical concerns. One study did a content analysis of local pandemic plans and found a striking lack of ethical language and direction.[32] The relative absence of ethical language should be a concern. Why? Because when a pandemic strikes and spreads quickly across the nation, there will be little time to reflect on ethical concerns. It will be impossible to adjust public health, medical, and response systems on the run to enable them to act ethically or make ethical decisions. Ethical issues such as those associated with the allocation of scarce resources or the constraints on civil liberties that may be imposed by various control measures cannot reasonably be addressed in the midst of a crisis. They require much advance consideration and planning. Some of the local plans alluded to ethical issues, but they did not flesh them out or resolve the concerns that were raised.[33]

States and localities did make progress in pandemic planning in relation to the H5N1 virus, but there remain obvious concerns that should be revisited, and planning should continue to improve our response. No matter how good the original pandemic preparedness plan, there is always a need to improve coordination between the public and private sectors, the monitoring or public health surveillance of the human and animal populations, and the risk communication and public education so as to prepare the general public. Also, there is always a need to anticipate the ethical dilemmas that a pandemic may present. In other words, as is always the case, plans are nothing, but ongoing planning is everything.

We know that counting on good luck is not a preparedness strategy. We know that some considerable effort has been expended on planning and preparing for an influenza pandemic. We have provided an overview of the H5N1 planning at the local level. For any of these local plans to be effective, it is imperative that the U.S. federal government meet some serious commitments at its level. A federal commitment to research, the development and stockpiling of vaccines and medications, and the efficient support required for the implementation of countermeasures are all critical to ensure that the nation can respond to a pandemic outbreak.

The most important federal commitment relates to the production of vaccines. The pharmaceutical market is a trillion-dollar enterprise, but vaccines make up only 3 percent of the market.[34] The pharmaceuticals do a reasonable job with seasonal flu vaccines, but, remember, the influenza virus mutates constantly. This means that new versions of the seasonal vaccines must be developed and made each year as these mutations occur. This is a process that takes many months and would be too slow during a severe influenza pandemic. It can require years of research and testing and over a billion dollars to develop a single vaccine. Drug companies find it too costly and unprofitable to make and store pandemic vaccines. The vaccines and other countermeasures needed to respond to a pandemic are thus largely dependent on federal funding for research, development, and manufacturing. Absent this, the private sector simply is not incentivized to meet the need. A public-private partnership under the Biomedical Advanced Research and Development Authority's (BARDA) pan flu program contracts with the private sector to develop pan flu countermeasures.[35]

Influenza-vaccine manufacturers cannot sustain the capacities to respond to a flu pandemic on their own. In the aftermath of the 2005–2006 pandemic planning in relation to H5N1, the federal government stepped up its efforts. From 2006 to 2013, federal funding for pandemic response was enhanced. It allocated $5.6 billion to implement the first two years of the pandemic influenza plan. When there was an outbreak of the H1N1 strain in 2009, an emergency supplemental appropriation of

$6.1 billion was allocated, but these funds have since dried up. The federal willingness to invest the necessary dollars for pandemic preparedness is in decline. BARDA's pan-flu plan received only $115 million in 2014 and $72 million in 2015, for example. At that low level of funding, the Department of Health and Human Services (HHS) has concluded that BARDA will not be able to award new contracts that are required to maintain the existing stockpile of countermeasures. This, according to HHS, makes it impossible to mount an efficient and effective pandemic response.[36]

When all is said and done, just how well prepared for a pandemic are we? We must note that considerably more attention has been focused on protecting the public from terrorist attacks than from the far more likely and pervasive threat of pandemic influenza. Science has done its job of alerting us to the countless pathogens that have the potential to cause enormous and deadly harm to us. The scientific and policy communities have begun to take the threat of pandemics more seriously. As we have seen, that has involved new planning and the commitment of new public resources. But as is endemic to the relationship between science and politics generally, the policy makers still lag behind the science. This lag, despite any progress that has been made, holds the potential for disaster.

Are We Prepared for the Next Pandemic?

At the beginning of 2017, a new avian influenza strain reanimated global concerns about a possible pandemic. H7N9 spread across China, infecting mostly poultry, but it had begun to spread from chickens to humans. Almost 9 in 10 humans infected with it came down with pneumonia. Seventy-five percent of the infected humans experienced severe respiratory problems and ended up in intensive care; 41 percent died.[37]

The experts fear that if H7N9 begins to be transmitted from person to person, the result will be the next global pandemic. This news was undoubtedly no more than a blip in the newsfeed for the few who noticed it and of no concern to the many who had heard nothing about it. But for the experts who know, H7N9 is China's fifth and largest outbreak of avian influenza. According to the experts, at the beginning of 2017, this was the influenza strain with the greatest potential to cause a global pandemic,[38] but this new threat seems not to have caused much concern in U.S. political circles.

As 2017 began with the news of a heightened risk of a global pandemic, the new Trump administration had not yet appointed senior officials to head key federal agencies responsible for pandemic preparedness. This meant leadership vacancies in the Department of Health and Human Services, the National Institutes of Health (NIH), and the Centers for Disease Control and Prevention. In addition to these unfilled positions, these agencies were subjected to deep budget cuts under the proposed Trump budget for fiscal year 2018,[39] including about an 18 percent (5.8 billion dollars) reduction in the NIH biomedical research funding. The NIH plays an essential role in detecting outbreaks, supporting research to develop vaccines, conducting international health research, and containing any outbreak that should occur.[40] The point is, leadership in these agencies that are responsible for the nation's health security and the resources necessary to ensure that they can effectively produce vaccines and implement preventive measures to control an outbreak of H7N9 or any other flu virus was not a top priority as this new threat emerged. Such is, unfortunately, the rule and not the exception.

Even before 2017 and its new threat, the budget cuts, the lack of leadership, and so on, there was a growing consensus within the medical and scientific communities that the U.S. government needed to strengthen its commitment to pandemic influenza preparedness. While the nation had made progress over the past 10 years, as 2017 began there was considerable concern that too much remained to be done before we could be considered to be prepared for a pandemic. The needs articulated by the experts included more support for research and development of flu vaccines, therapeutics, and diagnostics; the replenishment of countermeasure stockpiles and a comprehensive evaluation of national preparedness; and, yes, more robust spending in support of pandemic preparedness.[41] But as one might expect and even rationalize, our elected policy makers are rarely eager to embrace this agenda. The fact that we might need to embrace such an agenda to be truly prepared for an influenza pandemic is something that has difficulty gaining acceptance in the daily minutia of our politics. But new disease outbreaks occur and remind us that pandemic planning is something we really should not slight.

Soon after President Obama took office in 2009, a new influenza outbreak was under way: the H1N1 flu. This strain had spread from the pig population to the human population and had its roots in Southeast Asia, but the first known human cases occurred in Mexico, with cases popping up in the United States shortly afterward. What unfolded is what one would call, perhaps, a mild pandemic. When this new flu strain jumped from pigs to humans, it ultimately killed an estimated 203,000 people worldwide. As it spread, efforts were made to accelerate the production of a vaccine, but the first doses would not be available for 26 weeks. It would have taken a full year to produce enough for every American.[42] Thankfully, the H1N1 outbreak was not that severe. The virus first spread widely in late April and May, slowed but still persisted through midsummer only to heat up again in late summer as children returned to school, then peaked in late October and early November before it waned fairly quickly. A feared third wave never materialized in the winter. Ultimately, the death rate was lower than originally projected, but the number of H1N1 cases, hospitalizations, and deaths were nonetheless substantial.[43]

Early in the H1N1 pandemic, no vaccine was available. Most Americans were nevertheless quick to adopt two recommended central public health strategies. In the pandemic's first weeks, almost two-thirds of Americans said that they and members of their family had begun to wash their hands or clean them with sanitizer more frequently. Most had made preparations to stay at home if they or a family member got sick. Almost 4 in 10 also said they avoided exposure to others with influenza-like symptoms.[44] People also followed the media reports on progress in vaccine development. When a limited amount of vaccine became available, the public was divided over whether they would get vaccinated. Fears about the safety of the new vaccine discouraged many. People appeared to think there was a trade-off between accepting the perceived risk associated with the illness and accepting any perceived risk associated with the vaccine. Another reason for avoiding the H1N1 vaccine was the belief that it was not needed, especially as the threat seemed to wane over time.[45] Thankfully, the crisis was of short duration and the impact less deadly than it might have been in a worst-case scenario. Public confidence or perception of need aside, vaccine production during the H1N1 outbreak was inefficient and sorely lacking. It was difficult to get it to those most vulnerable in an efficient, timely manner. Bottlenecks in flu-vaccine production also slowed the process for getting the new vaccine out to the marketplace.[46] In a much more serious pandemic, this would have increased the death toll significantly.

Another problem that emerged during the H1N1 outbreak was the slowness of local health departments to alert the public to the health threat it posed. Only one-third of the 153 local health departments surveyed in a Rand Corporation study had posted information about the new swine flu on their Web site within the first 24 hours after federal health officials declared a public health emergency. State health departments did better: 46 of 50 posted some information about the outbreak within 24 hours of the federal announcement. Communication is critically important in any efforts to combat a rapidly spreading infectious disease.[47]

A postevent review of the U.S. federal H1N1 pandemic response by the U.S. Government Accountability Office (GAO) concluded that "public surveys generally found CDC's communication efforts to be successful in reaching a range of audiences; however, these messages fell short in meeting the needs of some non-English-speaking populations."[48] The deployment of the Strategic National Stockpile (a supply of medicines and medical supplies to be used for a national emergency) met the established goal according to the GAO review. However, CDC and state officials identified gaps in planning, including disparities between the materials expected and those delivered, as well as the need for improved long-term storage plans for stockpile materials. The credibility of all levels of government was said to have been diminished when the amount of vaccine available to the public in October 2009 did not meet expectations set by federal officials. While prior pandemic planning was helpful, it did not prove to be sufficient to meet all the challenges of the H1N1 pandemic.[49]

The H1N1 outbreak, which was declared a pandemic by the WHO, came at a vulnerable time for cash-strapped state and local public health departments. It followed a decision by Congress to cut $870 million slated for flu preparedness from the economic stimulus bill proposed by the Obama administration in the spring of 2009.[50] Republicans in Congress disagreed with many aspects of the Obama stimulus package. Understandably perhaps in the context of the economic crisis and the fierce partisan debate of the moment, cutting funding for pandemic flu preparedness was easily targeted. Many Republican members of Congress expressed the notion that spending for flu preparedness seemed to be a luxury that was not affordable at the time.

Spending money to prepare for something that *might* happen, to address the risk of a crisis that scientists tell us *might* occur at some uncertain future time, is a hard sell in the political universe. A review of the history of our pandemic and public health preparedness efforts makes it abundantly clear that the U.S. government does not spend in a way that makes preparing for a future pandemic a national priority. This is not to say that Congress does not spend money on national health concerns, rather that money gets allocated on a disease-by-disease basis and usually much too long *after* a crisis has begun. The approach is reactive as opposed to proactive. During the 2013–2016 Ebola crisis, Congress appropriated over $5 billion in emergency spending to address the health emergency, but this funding came about five months after international health groups had identified it as a crisis.[51] That reactive approach, and that sort of delay, simply will not be adequate to respond to a global influenza pandemic. Consider the congressional response to the Zika crisis. It took *nine months into the crisis* for Congress to allocate a little over $1 billion to fight a disease that was already spreading in the United States.

Some of that money came from or was diverted from existing Ebola funding. Scientists will alert us to the threats, and experts will say that the United States needs sustained spending for pandemic preparedness that extends for years, but what we will get is what we have seen to date.

The Ebola epidemic of 2013–2016 demonstrated just how ill prepared and clumsy the U.S. response can be to a global medical emergency. Ebola, or Ebola hemorrhagic fever, is a rare and deadly disease caused by infection with one of the five Ebola virus species, four of which are known to cause disease in humans. Scientists and medical researchers have known about Ebola for quite some time. It was first discovered in 1976 near the Ebola River in what is now the Democratic Republic of the Congo.[52] Since then, outbreaks have appeared sporadically in Africa. As a new outbreak took flight in 2013, the United States had already failed. It had neglected the disease and had not made vaccines and drugs to treat Ebola. It had also made cuts in the funding of essential public health agencies that would be needed to respond to any global medical emergency.

The worldwide response to the new Ebola outbreak in West Africa in 2013 was too slow. The WHO admitted as much and said that it had come up short with respect to the resources needed to respond. Charities such as Doctors Without Borders struggled to provide whatever assistance they could as the epidemic began to spread, quickly overwhelming the fragile health systems of poor West African nations. It killed thousands of people. To make matters worse, conditions generally in West Africa, including extreme poverty, dysfunctional health care systems, distrust of government after years of armed conflict, and the months-long delay in response, ensured that the death toll would mount.

The global response to the Ebola epidemic also highlighted major inadequacies in our ability to respond to public health emergencies in the United States. It was not until a few Ebola patients landed on American soil that the United States mobilized its efforts. In West Africa, more than 23,000 people had already been infected, and 9,600 had died. In the United States, eight patients from the region landed on American soil and two of them died. Also, generating much publicity, two nurses who treated one patient were infected. They both received treatment and recovered. Public health experts in the United States emphasized that the risk of catching Ebola from travelers was almost nonexistent, and the treatment of the disease in the United States was not nearly as uncertain as it was in poorer third-world nations. In essence, the risk to Americans was never very great. What was needed, they stressed, was U.S. and global assistance in combating the disease in West Africa. But most Americans were focused on their own almost-nonexistent risk of catching Ebola from travelers instead of the pressing need to help the truly affected nations. Indeed, the word used by a presidential commission that assessed the American public's reaction was "panic."[53] It was a panic wholly uncalled for and completely out of proportion with the reality of the situation.

One example of the panic that was bursting out across the nation was best embodied in a series of 2014 tweets by citizen Donald J. Trump. In one, he called for American health care workers who had been infected with Ebola while providing medical assistance in West Africa to be barred from returning to the United States: "The U.S. cannot allow EBOLA infected people back. People that go to far away places to help out are great—but must suffer the consequences." He also warned, contrary to all the legitimate evidence and expert analysis, that "Ebola is much easier to transmit than the CDC and government representatives are admitting."[54] Wild claims posted on Twitter during a global medical emergency only produce unnecessary public concern, possibly panic. This is the last thing one wants to see during a global pandemic scenario. Public confidence in government is essential for public safety in such a situation. Likewise, it is incredibly important that scientific and medical experts are heard and understood during such emergencies. The last thing that is needed is uninformed, evidence-challenged, emotional outbursts of misinformation and "alternative facts" that just are not true. In fact, many emergency response experts today are worried that the emerging climate in the United States, with its talk of "fake news" and "alternative facts," would work against any chance of a successful response to a future pandemic.[55]

The U.S. response to the recent outbreak of Zika is another reason to be concerned about the nation's preparedness for a major pandemic. By the spring of 2016, there were almost 2,700 travel-associated cases of Zika in the continental United States. Things were even worse in the U.S. territories (more than 14,000 locally acquired cases had been reported). As the crisis unfolded, Congress was dreadfully slow to respond. In the summer of 2016, they had yet to pass a funding bill. This presented the Obama administration with a difficult challenge to initiate a response without new resources. They decided to redirect money earmarked for other purposes, including Ebola response, to support Zika research and response efforts. Experts pointedly noted that the response to the Zika crisis, like the response to the Ebola epidemic, was shaped and thwarted by a fragmented and partisan U.S. political system[56] that was simply too slow and inadequate in the face of an emergency. It was not a response that served the public interest in the least.

In early February 2016, President Obama sent a proposal to Congress. He requested $1.9 billion in supplemental funding to strengthen local public health responses to the Zika threat. Included in this request was funding for hard-hit U.S. territories, such as Puerto Rico. Other major priorities included expanded Zika testing in the United States and support for research on a new vaccine. Given the global nature of the Zika threat, another $400 million was earmarked to fight the disease abroad. The House and the Senate responded with alternative bills. The Senate bill cut the requested allocations and opted for the reallocation of existing Ebola funds toward Zika response. The House bill simply offered one-third of the funding the administration requested. Naturally, things stalled in Congress. With no legislation forthcoming by the end of August, President Obama had no choice but to use his administrative authority to move $81 million already allocated for biomedical research, Ebola response, and other health programs. These funds were reassigned to keep a Zika vaccine study going and support the local governments[57] in their response to the outbreak.

Every dollar that would be spent from the spring to the fall of 2016 on the Zika outbreak, a time regarded as critical in responding to the threat, was money that the administration reallocated from other uses to combat the virus. This money would soon run out. Public health experts called the response insufficient and warned that moves to shift money away from existing public health initiatives, including Ebola prevention, would undermine national and local efforts to prevent other disease outbreaks. By October 2016, Congress had finally provided $1.1 billion in federal funds to combat the threat posed by the Zika virus. A portion of this was money taken from Ebola funding, of course. Public health officials were happy to finally receive the funding, but they vented frustration at Congress for taking so long to make the money for Zika available. By October 2016, the virus had spread to more than 25,000 people in U.S. states and territories, including 3,600 on the mainland. There is no doubt that such a delayed performance by Congress in a pandemic scenario would result in a tragic increase in the death toll.

The scattershot way the United States deals with pandemic or disease response is, given the nature of the threat, irrational at best and a disaster in the making at worst. What is needed is sustained funding and persistent planning that extends over many years. It is clear that the United States has worked to do some serious pandemic planning, and it is imperative to do all that we can to contain the threat of a pandemic. It is equally clear that if efforts to contain an outbreak at its source should fail, the resources of the federal government will not be sufficient to prevent or control the spread of a pandemic across the nation. Policy makers' ability to respond quickly, effectively, and intelligently in such an event is unproven at the moment. Pandemic preparedness must, as a logical necessity, also be aimed at addressing the resulting impacts on communities, workplaces, families, and individuals. As we conclude this overview of our national efforts at pandemic preparedness, there is more reason for concern than for confidence.

Conclusion: Panic Prone and Underprepared

Infectious disease experts have, over the last couple of decades, consistently urged the U.S. government to do more to keep the nation prepared for outbreaks of diseases. They have been especially insistent in their warnings regarding the need to prepare for the next influenza pandemic. Influenza specialists have repeatedly warned that a global influenza pandemic is a certainty. Flu, of course, mutates constantly. A major new strain emerges every couple of decades. The United States needs to be prepared ahead of time. This preparation includes, as we have seen, the development and stockpiling of vaccines, countermeasures, equipment, and plans for deploying them. It includes local preparedness, private-sector preparedness, and citizen preparedness to respond effectively when a medical disaster hits. Waiting until a pandemic is under way is too late to do any of the things required to be truly prepared.

One would think that protecting the United States from the next pandemic or killer flu would be something all policy makers could agree on. A pandemic is neither a Democratic nor Republican issue; it is a national concern that impacts the health and safety of all citizens. Be that as it may, the money is not there, and the effort has been inconsistent. As we have seen in our discussion of the H1N1 outbreak, the Ebola crisis, and the Zika outbreak, the national response is a bit too reactive and much too slow. Congressional funding is provided on a disease-by-disease basis, and the need to go to Congress for such funding needlessly slows the national response. This alone can be devastating in a rapidly moving pandemic scenario. As we saw with the H1N1 outbreak, Congress actually cut funding for pandemic preparedness as the H1N1 pandemic began in 2009. Congressional funding to combat H1N1 would not come until a full five months *after* it had been declared a health emergency. It took eight months for Congress to provide funding for the Zika outbreak. Ebola was totally ignored until it came to the United States months after the crisis began in West Africa. Thankfully, none of these threats were as severe as a major pandemic would have been, but it is not reassuring to note that we were unprepared to respond efficiently to any of these threats, and to make matters worse, policy makers took many months to provide the necessary funding. In a severe pandemic, this sort of response and delay in providing what was needed would have resulted in an unthinkable number of unnecessary deaths.

The H5N1 threat in 2005–2006 resulted in billions of new dollars dedicated to pandemic preparedness. In the aftermath of the H1N1 outbreak, the Ebola crisis, and the Zika crisis, billions of dollars in emergency funding was approved for each. After-the-fact funding is not an optimal approach, of course, but at least it was something. Ironically, as a new H7N9 flu strain emerged in 2017, and with it a new pandemic threat, the 2017 federal budget provided just $57 million to prepare for an influenza pandemic.[58] By all appearances, the U.S. government was backing off on preparing for the next big pandemic as a serious new threat was emerging. This endangers citizens, as it suggests a future that is more threatening and dangerous than it needs to be. While many may believe this to be alarmist or farfetched, we should remember what the researchers and public health experts are telling us: *There will be a next pandemic.* It does not take much to imagine a new deadly virus that spreads easily from person to person in a country that sends people around the globe every day. In that single day, the next global pandemic could begin. A pandemic is not really something to react to when it happens. It is something to prepare for well ahead of time.

Few things kill as many people as an influenza pandemic. One would think that the case for pandemic preparedness is a compelling one and that the first line of defense against any deadly outbreak is to finance pandemic preparedness at a national level. It is sadly the case that it is always easier to find the money to respond to an outbreak than to secure investment to stop one from happening in the first place. Most experts agree that reinforcing capabilities such as disease surveillance, diagnostic laboratories, and infection control would be far more effective and cost far less than trying to contain outbreaks once they occur. Yet the United States has not invested nearly enough in pandemic preparedness. For that matter, the global community has not invested nearly enough in preparedness. It is worth taking a moment to discuss why this is the case.

Human beings generally do not focus on the long haul; instead, there is a tendency, especially among Americans, to focus on immediate events. This, combined with a short attention span that seems to become shorter all the time in the minutia of our fast-paced digital age, means that the current situation (whatever it is today) influences our thinking more than the future. An influenza pandemic is tomorrow. In fact, it is probably a tomorrow that cannot be predicted precisely. Immediate concerns carry more weight in our thinking and conversations than long-term concerns or objectives. It is not that we absolutely deny such things as a possible new influenza pandemic. Even those who do not know everything that the experts are saying or warning about realize that new diseases happen, and rather frequently at that. They just do not believe these diseases will happen to them.

Humans are generally tuned in to extraordinary events with shocking visual imagery of unexpected or unusual intensity. Dramatic storms, fires, droughts, floods, and weather extremes will capture our attention because they contrast with what we regard as normal or expected. The same is the case with global health emergencies. We don't think about them until we see the pictures or the television reports that bring them into our homes. But even then, what we are seeing is a sad thing for somebody else but not something we expect to happen to us. Nobody expects to be flooded out of their house when they watch a news story about somebody else being flooded out of theirs. But when these things happen to us, especially when a serious new virus is in our community and is a threat to our family, we have a sense of urgency about the virus and about our safety. That's pretty much how most lives go as they are lived out in everyday United States. This is why we need scientists and medical research. This is why we need public officials and policy makers to digest what the scientists and researchers tell us and make the decisions necessary to see that we are prepared for major health and pandemic threats. But in practice, and as we have seen, policy makers are generally not preparing us to the extent that the experts believe they should. Indeed, the experts are telling us that we are not prepared for the next major pandemic.

Our policy makers, our politicians, are also focused on more immediate events. They deal with the political minutia of the moment. The ideological battle and the partisan fight to control the direction of economic, social, security, and other policy initiatives consume their attention and time. Thinking about the next pandemic is lower on their list of priorities. This is not to suggest that they are inattentive to such public health threats, only that they are not really motivated to act with any urgency until the crisis is unfolding or at hand. Even then, as we have seen with H1N1, Ebola, and Zika, they move slowly, slowed down by their ideological or partisan disagreements. The relationship between science and politics here is much different from the collaborative dynamic, the conflict dynamic, and the resistance dynamic. This panic dynamic is one that sees the science working hard to catch up with events and to explain what is happening. Indeed science is incredible in its ability to warn us about public health emergencies. But until such an emergency unfolds, the public is detached or uninterested, and the policy maker is slow to engage. When a pandemic or public health emergency unfolds, the public is in a state of confusion or panic, and policy makers are without immediate answers and take a considerable amount of time to gear up and respond. Fear may grip the public, and politicians are initially flummoxed as scientists seek to project a sense of calm and inform an uneasy public.

The Ebola crisis of 2013–2016 is a perfect example of the general preparedness problem. Let's recap what we have already said about that. Scientists and researchers had known about Ebola since 1976. They knew just about all that was knowable about it. They were ready to explain the new outbreak in West Africa, and they informed us of how dangerous that could be in an undeveloped and impoverished part of the world where treatment and facilities were sorely wanting. But despite all that they knew, the world was slow to react. The World Health Organization was slow to react. The United States was slow to react, so much so that it was deemed to be unprepared to address a global medical emergency. It took months after the crisis was identified for Congress to act. As a few cases arrived on American soil, including infected health workers, the public exhibited signs of alarm. While health experts worked hard to inform the public, the public reaction lurched toward an unnecessary panic. The experts particularly wanted to make sure that the public understood that the risk of catching Ebola from travelers or returning health aid workers was almost nonexistent, but the main public reaction was a fear-driven panic reflex about what they were clearly being told was a nonexistent threat. This unfounded panic even included, as we saw, a tweet storm from a source that would become an all-too-familiar and dangerously uninformed serial Twitter stalker of the nation in the years to come.

Science had done its work, and the experts knew what was happening and what would happen as the Ebola crisis unfolded. The policy makers were slow to react, not because they were reluctant to follow the advice of science but because the dynamic was generically unhealthy. It was not conflict or resistance at play here; it was simple inaction due to the reactive policy makers' impulses not being compatible with the anticipatory or predictive capacities of science. It was due to the fact that science and politics often do not engage each other most productively on such issues until *after* the crisis is unfolding. In other words, the panic reflex or the crisis was necessary to get the two on the same page. That is, as we have said all along, a prescription for disaster.

Victor Dzau, president of the U.S. National Academy of Medicine, speaking from the 2016 World Health Summit in Berlin, Germany, reported on the findings of an international study the U.S. National Academy of Medicine had been asked to implement. Over a six-month period, the academy, in consultation with hundreds of experts, examined current disease response preparedness and projected future global response preparedness. The study concluded that the world is nowhere near prepared for a major pandemic. It noted failures at all levels: "At the international level, there is a lack of coordination and resources; at the national level, there is a lack of public health infrastructure, capacity, and workforce; and at the local level, there is a lack of community trust and engagement."[59]

Within the United States, there is a growing awareness that the threat of a global pandemic is heating up rapidly at a time when the U.S. government is not adequately prepared to respond to one. General Mark Welsh, the U.S. Air Force chief of staff from 2012 to 2016, oversaw the military's contributions in response to the Ebola epidemic as a member of the Joint Chiefs of Staff. As that epidemic peaked and threatened to spread across the globe, Welsh observed a lack of planning and overarching leadership. This casts great doubt, he concluded, about the U.S. government's ability to respond to a global medical crisis. The U.S. government simply cannot be expected to overcome a lack of planning and to deploy experts and support staff to affected areas in an effective manner so as to contain and mitigate the spread of a deadly virus.[60]

Thinking purely domestically, misinformation about and resistance to vaccines, the absence of a clear plan for coordination among federal agencies, and a need to improve public awareness about the threat posed by a biological outbreak are among the factors that make the possibility of large and deadly pandemics increasingly likely. Take vaccines as an example. The antivaccine movement in the United States, which includes people who believe in the debunked link between vaccines and autism, contributes to the threat of a pandemic. As fewer children receive the childhood vaccinations because of this unnecessary panic reflex that has gripped their parents, we run the risk of falling below what is called "herd immunity." Herd immunity is achieved when the number of vaccinated people in a particular community is enough to provide protection for those who have not been vaccinated. Studies show that public-school children in states like Washington and Texas are already in danger of falling below this threshold.[61] It is not helpful when elected officials, such as the president of the United States for instance, openly question the science behind vaccines. Sadly, this was the case in 2017. It is not helpful when, in the midst of a pandemic, the public does not have confidence in the new vaccine developed to prevent its spread. Sadly, this was the case during the H1N1 pandemic. It is never helpful to be unprepared to respond effectively when a pandemic unfolds. Sadly, this is something the experts are presently warning us about and with great urgency.

Any objective assessment of infectious-disease outbreaks and pandemics tells us that the United States and the world need to recognize that we are unprepared to respond to a global crisis. All the informed sources tell us that the nation and the world need to be prepared through enhanced coordination, more investment, and better infrastructure, such as improved systems for disease surveillance. One suspects that most policy makers generally understand the importance of all this. But the fact remains that the science and the politics of the matter have not interacted smoothly to address the widely perceived and broadly shared need to be prepared. This is not a matter of conflict or resistance; it is primarily a matter of what we might call a collision of worlds that is endemic and very difficult to avoid. We must always remember that scientists and politicians live and work in different worlds.

Recall something we said in chapter 1: at its most basic level, science is the pursuit of knowledge and its application to the natural and social world. It seeks the truth about the physical world, and it follows a systematic methodology based on strict rules of evidence. Politics, at its most basic level in the United States, is the process through which differing interests compete for advantage with respect to the formal allocation of values, benefits, and costs in our society. We said that scientists work in a world where the result of their work is independent of social nuance. The aim is to remove context from the work. The results must be as broadly generalizable as possible. Politicians work in a world where context is everything. In other words, politicians are reactive. They respond to the most immediate and pressing problems, to their constituents, to special interests, and to the partisan values that motivate them. They are focused primarily on the minutia of the moment. Scientists deal with the grand questions of the universe. They seek to look forward, to expand their capacity to explain and to predict. At its best, science is proactive in exploring, discovering, and changing the very way we think about the universe and everything in it. All this means that scientists and politicians think and communicate differently about everything. In many ways, this makes them incompatible when it comes to the identification of priorities, problems, and solutions. It takes considerable effort to get them to address the same reality. Even when they agree on something, we have what might be called the Cool Hand Luke syndrome at work: "What we've got here is failure to communicate."

Even under the best of circumstances, politics is politics, and it is all about winning a partisan argument or serving a particular constituency or interest. This basic reality impedes, or at least makes inefficient, the process of cooperation in the anticipation of and response to a crisis. In the present iteration of political life in the United States, all facts have become a matter of partisan dispute. Bipartisan agreement is something so out of date that literally no living American can remember a time when there was any of it to be observed. When every fact is disputed, and when "alternative facts" or untruths are as valid as scientifically proven conclusions, it complicates the process of preparing for the next pandemic. A pandemic is not a partisan issue. Everyone agrees. But our political world is simply not able to react as efficiently and with the urgency that the scientific world says we must. This creates stress and an unhealthy global reaction. It leaves the world with a bad case of panic reflex. It may be time to try, however long the odds against success might be, to address the Cool Hand Luke syndrome.

CHAPTER SEVEN

Two Worlds—One Reality:
A Path Forward?

K. T. McFarland, the deputy national security advisor, entered the Oval Office and handed the president of the United States a printout of two *Time* magazine covers. One was supposedly from the 1970s. It warned of a coming ice age. The other, from 2008, was about surviving the challenges of global warming. President Donald J. Trump was soon amped up and ready to tweet something nasty about the "fake media's" hypocrisy. In this instance, unlike so many others in his early days as president, staff members chased down the truth and intervened before the president tweeted or talked publicly about it. It turned out that the 1970s cover was a fake that had been circulating for quite some time. It was an Internet hoax that had been fooling people for years.[1]

McFarland, a former *Fox News* analyst, had been appointed deputy national security advisor by President Trump. Soon after this incident, although completely unrelated to it, and after the firing of her boss (national security advisor Michael Flynn), she was demoted and named ambassador to Singapore. That she found the myth about scientists predicting global cooling in the 1970s convincing, and that she shared it with the president, is relatively insignificant. What is significant is that the president of the United States was taken in by the fake cover. He fell for it. More significant still is that this president believed that the 1970s cover was accurate, even after it was proven to be fake. As one White House aide supposedly said, it was "fake but accurate."[2] The president, and most of the people around him for that matter, believed it was true that there was a period in the 1970s when scientists were predicting a new ice age. In truth, this assertion is one of the oldest pieces of absolute nonsense that is still circulating on the Internet. It is a falsehood embraced by those who deny the reality of global climate change. There were a few articles in the popular press at the time hyping a coming ice age, but these articles were not works of science or suggestive of anything close to a scientific consensus. They never even really said or meant what the climate skeptics or deniers think they did. Some journalists had plucked the results of some exploratory and very incomplete work that speculated about the cooling effects of aerosols in contrast to the warming effects of natural gasses and wrote a few creatively exaggerated articles that overhyped the possibility of global cooling. But there was *never* a consensus or any serious and proven science that predicted a long period of cooling lay ahead.

Think about this for a moment. The president of the United States is shown a fake magazine cover and falls for it without question. Once the fake is pointed out, he is nevertheless convinced and reassured by his advisors that while the cover is a fake, the fake is accurate. In other words, the false and long-ago-disproven assertion that scientists were once predicting global cooling with the same certainty that they now speak about global warming was believed. The president, believing and being advised that the fake was accurate, was taken in by one of the biggest hoaxes ever perpetrated by the climate-change deniers in their attacks on legitimate science. This sort of thing happens in our public discourse when science and politics collide. But should it be happening in the White House? The question of policy preferences and partisan values aside, what level of knowledge and awareness should citizens expect and demand from the Oval Office? It is more than a bit disconcerting that any president could be so easily taken. He fell for a phony version of something even phonier. Scarier still, nobody around him intervened with any scientifically accurate information. What does this say about the relationship between science and politics in the early 21st century?

As the 45th president of the United States took office in January 2017, he demonstrated with swiftness and precision that science would be treated as an obstacle to be removed by his administration. The very moment he took the oath of office, all mention of climate change was removed from the White House web page. One agency in particular was singled out. By day three of the remarkable and very exhausting misadventure known as the Trump administration, he barred employees of the Environmental Protection Agency (EPA) from posting updates on social media or providing information to reporters. A few days later, the new administration mandated that all EPA studies be first reviewed by political staffers before being released to the public. This policy more or less reversed an Obama administration rule that had allowed scientists to work uncompromised by political or other interference.[3]

By March 2017, the Trump administration proposed cutting the EPA budget by 25 percent. This signaled a major scaleback in the EPA's role of protecting the environment. The EPA would also, under the direction of the new administration, withdraw an Obama-era request that oil and natural gas companies provide information on their methane emissions. Scott Pruitt, the newly appointed and confirmed EPA administrator, proclaimed in March that he didn't believe carbon dioxide is a "'primary contributor' to global warming."[4] By April, under Pruitt's leadership, the EPA scuttled efforts to clean up coal power plant pollution. As frosting on the cake, industries would no longer be required to reduce emissions of toxic chemicals. This was part of a broader set of moves by various Trump appointees to help companies that profit from burning fossil fuels. In May, Pruitt fired half of the 18-member Board of Scientific Counselors, a panel that evaluates research done by EPA scientists. This panel was created to help government regulators create the rules that protect clean air, water, and soil, among many other things.[5] The Republican Congress was also helping to keep science in its place. As an example, a proposed House bill would require more industry representatives on another EPA science panel so that its oversight of industry would be more "balanced" toward industry. On June 1, 2017, the Trump resistance to or denial of science inevitably resulted in the withdrawal of the United States from the historic Paris Climate Accord.

The Trump White House was not the first new administration to attempt to resist, control, influence, or ignore scientific research that contradicts its policy ambitions or the views of its political party. Presidents in both parties had previously ignored or sought to control science when it proved to be politically inconvenient or undesirable for them. Under President George W. Bush, Interior Department administrative officials sometimes overruled agency scientists working on endangered species issues, and scientists accused the Obama administration of underreporting the damage from the 2010 BP oil spill in the Gulf of Mexico. But the speed and intensity with which new controls were clamped on science in the Trump administration was unprecedented.[6] This caused an immediate reaction and not only within the scientific community. Unofficial Twitter accounts were launched to voice resistance to the Trump administration's quick succession of orders. Some claimed to be tweeting on behalf of unidentified federal scientists. Resistance within the federal government was rumored at the Park Service, NASA, the U.S. Forest Service, the EPA, and the Agriculture and Health and Human Services Department. By April there was a March for Science, inspired no doubt by the historic Women's March on Washington, which drew about a half million people to the nation's capital the day after President Trump was inaugurated. Scientists, of course, are not known for waging political protest en masse. But many of them, and a large segment of the public, were willing to do just that in response to what they perceived as unprecedented attacks on science from the new administration.[7]

The new Trump administration in 2017 was really *not* the problem. It was the inevitable result of a problem that has persisted for many decades and that had reached a new level of seriousness in the first couple of decades of the 21st century. Whether or not the Trump administration is long-lived, and there was from the very beginning some doubt about its life expectancy, as controversies very quickly ripened into questions of legal and constitutional importance, is irrelevant. The Trump administration does serve to remind us, however, that we have ended up in a very strange and troubled place in the relationship between science and politics. As the previous chapters have suggested, this places serious, perhaps even existential, challenges before the current generation. The manner in which we will deal with these challenges may very well determine not only the conditions facing future generations but whether or not there will be future generations. This makes the present moment and its politics more interesting to be sure, and the new kinds of citizen activism that are emerging in response to the Trump administration are fascinating to observe. But this book is focused on the relationship between science and politics in the shaping of our public policy. While the current drama in this relationship plays itself out in our political life, we must take a broader view of the problem and give some serious thought to what is required in terms of possible solutions. It is to these tasks that we will devote our attention next.

Most of what follows reflects this author's perspective and thinking. Not all of it is original thinking, of course, and there are many similar discussions taking place in both the scientific and political worlds. The definition of the problem and the discussion of solutions will provide a helpful framework for further discussion, but it cannot hope to be more than a broad outline for a dialogue that must include many more voices in a national discussion. The preceding chapters are intended to provide a basis for recognizing a problem and providing the outline of what will be necessary to address it. When science and politics collide, the results, as we have seen, are most often not in the public's best interests. But as we have also seen, addressing that reality is a monumental challenge. We must nevertheless seek to understand and meet that challenge.

The Problem

Before defining the problem, let's take a few moments to reflect on the culture that has given birth to it. It is, after all, the culture that produced a president who was taken in by a climate change denial hoax. In 1963, Richard Hofstadter published his landmark book *Anti-intellectualism in American Life*, which won the 1964 Pulitzer Prize for General Nonfiction. Hofstadter traced the social movements that altered the role of intellect in American society. He argued that both anti-intellectualism and utilitarianism were greatly enhanced by the democratization of knowledge. He did not see this as a good thing. The democratization of knowledge, as he saw it, altered the purpose of education and reshaped its form. Access to education became more important than excellence in education. Anti-intellectualism, according to Hofstadter, is a "resentment of the life of the mind. It is also a resentment of those who are considered to represent it."[8] There is, he argued, an American disposition to constantly minimize the value of that life. This is not to say that Americans are dumb. Hofstadter noted that there is great intelligence in Americans, just as there is great professionalism. The problem, as he saw it, is that professional intelligence is mechanical and functional. In other words, it is purely, and to the detriment of the nation, utilitarian.[9]

Anti-intellectualism and utilitarianism were, as Hofstadter saw it, functions of American cultural heritage. They were actually built into the very fabric of the nation. But they were definitely not good things for democracy. Knowledge, in the utilitarian view, is about the completion of tasks and the execution of a formula. The mind is seen as having the operative mode of a machine. The preferred way of exercising the mind, for many Americans, thus takes on what Hofstadter labeled "mediocre sameness."[10] There are only so many ways to perform a task, for example, and for Americans, that performance is generally meant to attain some material objective. Many Americans learn at a very young age that their entire life is about the job or the career they will one day have, and they think about it usually and primarily in terms of the monetary or material rewards it will bring them. This is how success is most frequently defined for them in the prevailing culture. Knowledge, such as it seems useful to them, is akin to assembly methods in an instructional manual. For too many Americans, knowledge is thus practical and of benefit only in relation to the goal of self-advancement. This gives rise to a "mystique of practicality," to use Hofstadter's words, that stupefies people into voluntarily enlisting into the "curious cult of practicality."[11] Today, universities are criticized because they require students to take courses in fields that are not "career oriented," such as philosophy, literature, and cultural studies. State legislators are ever more frequently saying that public universities should, as their primary objective, be preparing students for career success in the marketplace. They should also do it with less public funding, or so the argument goes. As Hofstadter wrote in his description of the "spiritually crippling" cult of practicality, "parents send their children to college for the gains measurable in cold cash which are supposedly attainable through vocational training."[12]

From Hofstadter to the present, American culture has been characterized by the accelerating and almost complete dismissal of science, the arts, and humanities as the measure for knowledge and quality. These things are no longer the foundation for opinion. They have been replaced by entertainment, self-righteousness, willful ignorance, and deliberate gullibility as the justification for all opinions. Accelerating this trend was the arrival of the instant gratification go-go-go digital age. According to Mark Bauerlein, author of *The Dumbest Generation: How the Digital Age Stupefies Young Americans and Jeopardizes Our Future*, the result is essentially a collective loss of context and history, a neglect of "enduring ideas and conflicts."[13] Bauerlein, a former director of research and analysis at the National Endowment for the Arts, acknowledges that the digital age does hold the potential to increase opportunities for education, cultural activity, and political activism. But, he contends, the much-ballyhooed advances of this brave new world have failed to materialize. In fact, he says, they have actually made us "dumber." As a result, U.S. youth know virtually nothing about history and politics, and they have developed a "brazen disregard of books and reading."[14]

In an age when we have more information at our fingertips than any previous generation, we also have an unprecedented lack of respect for knowledge-based and relevant information. There is a limitless supply of junk on the Internet and our social media, and much of it is designed to confuse and mislead us. In other words, much of the information available to us in the digital arena is simply noise. It competes with and frequently overwhelms knowledge-based or expert analysis that is also available online. Even quality education is undermined in this environment. Yes, we have online degree programs, but many of these are of dubious quality when they are packaged to "customers" rather than students and aimed at generating profit or enhancing enrollments, often at the expense of knowledge and the best pedagogical practices. Even those responsible for educating us often have a disdain for intellectual pursuits as they have enlisted full bore into the "cult of practicality." They have elevated utilitarian priorities over intellectual value in the programs they design and the product they deliver. This does not mean that customers (formerly known as students) will not acquire some marketable skill (although even that cannot be guaranteed) with their degree. But it does mean that they will find it possible to graduate with precious little knowledge about science, history, and politics, and they will still have a "brazen disregard of books and reading." While some might suggest we are creating a world of dummies, it is perhaps more accurate to say we are creating a world in which reasoned, informed, and intelligent thought is no longer possible. This is how a democracy starves to death. A democracy needs both an engaged and informed citizenry. Getting the basic facts right, understanding reality, is a prerequisite for democracy to work. Among the most critical tasks facing political leaders in a society as complex as ours is to create a consensus, a governing coalition if you will, among people and interests holding competing views. This is very difficult to do, even when everyone agrees on the underlying facts. When reasoned, informed, and intelligent thought is no longer possible, agreement on what is real (i.e., the underlying facts) is not possible. This is a potentially fatal condition for a political system that requires an informed citizenry. Public misperceptions replace knowledge and reason and become a formidable obstacle to a functioning democracy.

Valuing the work of intellectuals and seeing to it that we have a well-educated public with a basic common knowledge in the areas of science, math, history, philosophy, literature, and so on is absolutely essential for a successful democracy. An informed and competent citizenry is essential for a workable democracy. Reading books and knowing stuff matters. What American culture has produced as an alternative is not, at least by the measure of our interactions on social media, reassuring for those who value a successful democracy. What we see is a world of angry and under-informed people who cannot separate fact from fiction. They feel entitled to comment on everything, to make sure their voice is heard above every other voice, to personally attack all contrary voices, to dismiss and disrespect experts as elites or enemies of the people, to proclaim the facts to be whatever they believe them to be, and to view any proven facts that contradict their opinions to be part of a conspiracy. All this is done (even by the president of the United States) loudly, repeatedly, confrontationally, and stupidly. This results in the emergence of anti-intellectual celebrities and conspiracy gurus who become opinion leaders and the heroes of their many followers. It also results in a culture of antirationalism in which *every fact is suspect, every rational thought is the enemy's trick, and critical thinking is unholy and unpatriotic.*

Add to this description of American culture the evolution of "tribalism" or extreme partisan division in our politics. This too shapes and defines the problem that infects the relationship between science and politics in the contemporary United States. Political polarization has become the defining feature of early-21st-century American politics, both among the public and its elected officials. Republicans and Democrats are further apart ideologically than at any point in recent history. Recent surveys show that a growing number of Republicans and Democrats express highly negative views about the opposing party. They view each other as enemies. More Americans identify as more conservative or more liberal than in the past, and fewer say they are moderate or middle of the road. This means that partisan antipathy has risen, and it has made compromises across party lines more difficult.[15] "Ideological silos," socializing only with people who share one's own views, have become more common. The most ideologically oriented Americans, in both parties, make their voices heard through greater participation in every stage of the political process. To those on both the ideological right and left, compromise now means that their side must *always* get more of what it wants than the other side.[16]

In the current political climate, Democrats and Republicans agree on less and less, and that is not the biggest concern. On a more basic and often more tragic level, they do not accept the same facts as the starting points for debate. It is one thing to disagree and to debate over priorities and policy options; it is quite another to disagree and debate without a common or shared reality. Reality itself has become a matter of partisan perception uninfluenced by any objective foundation. Climate change is a scientific issue, for example, where Republicans and Democrats have opposing and irreconcilable views. In fact, as we saw in chapter 3, political ideology or partisan identification is the variable that shapes most of the public's views about climate change. The actual science of the matter was shown to have much less influence on the distribution of public opinion. Then U.S. senator Obama famously denied the intensity and bitterness of the partisan divide. In the 2004 convention speech that brought him to immediate national prominence, Obama argued that there was not a red United States, and there was not a blue United States, but there was one United States of America. It turns out he could not have been more wrong. There is a red United States. It is the sparsely populated but vast landscape of rural and suburban areas. There is a blue United States. It is found in the urban centers where single women, minorities, and cosmopolitans cluster. These two United States have less and less in common, and they find it increasingly difficult to communicate with each other. It has been obvious to everyone for some time that U.S. political life has become extremely polarized. Not only are policy makers further and further apart, but the American people generally are further and further apart, both geographically and ideologically.

Politically, we have become a nation largely sorted into two teams. The driving force in our politics is most frequently the contempt these two teams have for each other. We are a people less and less able to understand, communicate, or empathize with the other side. In fact, many observers are speaking about hyperpolarization as the new norm in American politics. Anger and divisiveness in American politics are not new things, but it has become so intense that it has produced an unhealthy anxiety in the culture. The polarization of our nation is, according to some, the product of conflicting and incompatible worldviews. Differences of opinion concerning the most provocative issues on the contemporary political agenda (e.g., race, gay marriage, illegal immigration, health care, the use of force to resolve security problems) have seen the left and the right coalescing around dramatically opposing and irreconcilable worldviews. This has preoccupied our politics with more uncompromising viewpoints more rigidly argued, which has proven to be more destructive of our capacity to govern than any functioning society could be expected to endure. It has even been argued that our democracy simply cannot work anymore. Authoritarianism has become an especially compelling alternative for some partisans who have come to feel vulnerable in the context of the uncertainty and conflict born of stubborn intractability in contemporary American politics.[17]

Whether discussing our culture or our political partisanship, these are the things that our representative government represents, along with the varied special interests of the groups that comprise the constituency of or pay to elect our policy makers. Knowledge is not a prerequisite for public service. Often, as Adlai Stevenson long ago so clearly demonstrated, it is a handicap. Some politicians, the really accomplished ones, have learned enough about our culture and our partisan behavior to manipulate both to their advantage. In dealing with the minutia of the moment, the main business of politics, they have developed the capacity to seem to be what the voters want even to the point of reinforcing the anti-intellectualism of the culture or the disruptive partisanship of the current age. Despite this, they still sometimes manage to govern and to accomplish a few constructive things for the people. But their number is dwindling. Other politicians, the really bad ones, are simply reflections of the hyperpartisan disruption that erodes our capacity to govern. They are the diseased product of the cultural and partisan garbage heap that has accumulated. They are the unmistakable manifestation of the rapid spreading of a political plague that has been unleashed in the absence of the sanitizing properties of knowledge. Their efforts produce nothing of public value and often succeed only in causing the paralysis of government and the abandonment of public policy as a tool to serve the broader public interest.

It is in this cultural and political environment that the relationship between science and politics, and the potential for science to influence public policy, must be evaluated. Our culture and its anti-intellectualism and utilitarianism have *always* guaranteed that science and politics would often be working at cross-purposes as they lived and operated in very different worlds. The emergence of hyperpartisanship in our politics has served only to separate the two even further and to substantially weaken the ability of science to be most effectively incorporated into the policy process. Far from being the ideal, the historically troubled relationship between science and politics has escalated into a political war on science that pits the power of science against the power of money. The power of science is losing whatever independent status and influence it once had in this competition, and its expertise is increasingly mistrusted and denied by a public that gets its news from Internet hoaxes. The power of money is winning.

Ideally, the disinterested and objective expert would be seen as the rational and authoritative arbiter of public disputes over scientific or technical issues. This is certainly a view that scientists, however reluctant they are to engage political matters, might promote when discussing the relationship of science to the policy process. This is an old notion. The appeal to facts and the interpretation of them by accredited experts is straightforwardly appealing. There are, of course, always limitations to the ability of experts and expert knowledge to influence public policy. Experts do not always agree. The public knows and is often exaggerated in its sense of how experts can and do disagree. Experts are not infallible, and even the most rigorous scientific methodology cannot guarantee complete objectivity. Expert advice might very well be influenced by professional or economic considerations. But beyond such limitations, it is important to note that public trust in experts, in the neutrality of expertise, and in the reliability of expert opinion is questionable in the current cultural and partisan environment. This complicates the incorporating of scientific input into technical decision making and policy formulation. Limitations can be identified and overcome, or at least managed. Mistrust is not so easily overcome or managed when it becomes the cultural norm. It is not a death blow, and it may be addressed in constructive ways, but mistrust is impossible to address without recognizing the major contributing factors to its existence. The major difficulty, and the major multiplier of negative effects in an already troubled relationship, is the fundamentally more dysfunctional relationship between science and politics spawned in our current cultural iteration.

The two major premises of this book are that (1) if scientists do not offer their expert knowledge to the policy process, or if policy makers do not incorporate this expertise into their analysis, we risk making ill-informed policy decisions on technically complex issues and (2) science and politics have an inherently troubled relationship.

The scientific and political worlds collide, to our disadvantage. As we seek to define the problem, we must understand both this troubled relationship and the culture in which it resides. We have, in the previous chapters, discussed the relationship in the context of four basic dynamics at work in it. In summarizing these dynamics here, and connecting them to the broader cultural and political trends currently at play, we will be able to see the problem to be solved more clearly.

The first dynamic in the relationship between science and politics that we assessed was the collaborative dynamic. "Collaborative dynamic" refers to a situation where political policy makers and scientists are in basic agreement as to the policy goals and the scientific knowledge necessary for achieving them. The agreement is such that the level of trust between the two is unquestioned. Likewise, relevant special interests and/or the general public seem to be in agreement with the policy goals. This situation is extremely rare, and it seems to be a dynamic that was politically as opposed to scientifically motivated. It was the Cold War, or a national-security crisis, that initially led to the commitment of tremendous political capital and governmental spending in support of a first-strike nuclear infrastructure. A significant portion of this spilled over to the scientific and aeronautical fields, which were associated with a peaceful and more optimistic message. The manned space program and the moon adventure were happy by-products of the Cold War. The science that was necessary to achieve national-security needs, and the manned space objectives that grew out of them, was not the driving force behind the collaborative dynamic; it was the means of implementing something that the political partner had already decided to do for reasons other than science.

The collaboration between science and politics during the Cold War resulted, as we said in chapter 2, in a marriage of convenience. From the beginning of the Manhattan Project to the landing of Americans on the lunar surface, this marriage flourished. But when the political priorities changed as the Cold War wound down, the marriage ended. Science had been an indispensable and inseparable mate to politics during the Cold War, but its future relationship with the policy process would be very different. Of necessity, there would be a relationship but on very different terms and with less respect for each other. The collaborative dynamic of the Cold War period has not been replicated since. Science has been useful to and used by policy makers since that time, of course, but the relationship has never again been a marriage; it has been more like friends with benefits. Grants for scientists, the generation of some sound policy advice on complex scientific and technological issues, and some piecemeal advances toward responsible public policy have been positive outcomes. The negative outcomes include long, never-ending disputes that have often complicated problems rather than solving them. This has been the too-frequent result of an uneven and often contentious relationship.

It must be noted that the collaboration between science and politics typical of the Cold War era, and the space age, was mechanical and functional. It was purely utilitarian. The political policy makers had identified a practical goal: national security required the development of first-strike nuclear infrastructure. Space exploration, as an outgrowth of that effort, was seen as a logical or functional extension of that practical goal. Achieving the advantage of the ultimate position in the arms race necessitated the mastery of space technology. Science was a functional necessity in all this, and its value in the partnership with politics was defined in terms of its usefulness in achieving important national security objectives. Scientific knowledge, in other words, was highly valued in the "cult of practicality" that dictated the course of the nation's national-security needs. The expertise of scientists was celebrated and respected primarily in relation to this utilitarian purpose. When that purpose had been achieved and other practical needs dictated other priorities, the value of the science and of the space program itself was quickly diminished. As we saw in chapter 2, shortly after the first moon landing, the public came to feel that the effort had not been worth the expense.

As the Cold War crisis began to abate, policy makers—like the general public—became more reluctant to continue funding lunar exploration, and NASA's mission was redefined. While not eliminated, the space program became less of a priority and would be maintained as a more limited enterprise defined by other practical considerations. The value of the space program was never really defined in terms of the scientific knowledge about the universe that it could generate but by the practical political objectives that the science could serve. As the Cold War ended, those practical objectives became more limited. Some of these were genuinely scientific, but they were deemed valuable for their practicality as much, if not more so, than for the sake of knowledge itself. In other words, the intelligence applied to space exploration remains mechanical, functional, and purely utilitarian for the policy makers and the general public. Even under the best of circumstances, even within a collaborative dynamic, science does not set the priorities or influence the direction of public policy; it is merely useful to policy makers who have already decided what they want to do. When science gets in the way of what the policy makers wish to do, they will not hesitate to fight or resist it.

The second dynamic we discussed, the conflict dynamic, has been more or less the norm since the end of the Cold War. The conflict dynamic is one in which we have an issue or concern where the conclusions or recommendations of science generate strong opposition from special interests and the political entities that they fund and support. Such opposition, as we have seen, is based most frequently on economic or material interests and becomes a part of the policy debate, often to the extent of muting or ignoring the actual science. In chapters 3 and 4, we discussed the conflict dynamic at considerable length in relation to climate change and hydraulic fracturing for natural gas.

Fossil fuel companies want to make money, and there are trillions of dollars still to be made. This is not necessarily a bad thing. These companies employ a good many people, provide shareholders a nice return on investment, supply the energy needs of our nation and the world, and it can even be argued that they have made life better by meeting these energy needs. But they also mine resources from the environment in an unsustainable manner; dump pollution back into the environment; shift the costs and ill effects of this pollution back onto the public and future generations; and are combative, duplicitous, and most often corrupt in opposing any and all governmental regulation of their activity for the broader public good.

In the case of climate change (chapter 3) and hydraulic fracturing (chapter 4), we saw that the fossil fuel industry and the political policy makers who are most supportive of their agenda are seemingly in a perpetual state of war with science. With respect to climate, the industry and its political supporters have long wanted to avoid acknowledging the basic inconvenient truths of science. With respect to hydraulic fracturing, they want to preemptively discredit or block any inconvenient truths that solid science discovers as it catches up in its assessment of the risks posed by the new technologies associated with fracking. In both cases, the political conversation is too often a fierce and very contentious debate between science and industry. In order to seek its profits and protect itself against governmental regulation, the fossil fuel industry and the policy makers who promote their interests find themselves denying or discrediting science to advance material interests. This has been quite successful, and the politicizing of science by its critics in today's hyperpartisan environment reduces scientists' ability to be heard above the noisy disinformation campaigns of industry and antiscience partisans.

In our discussion of climate change in chapter 3, we saw that the scientific consensus about climate change is based on a long history and an indisputable body of scientific work. We also saw that, as the science became more robust and certain, the fossil fuel industry worked to create doubt and uncertainty about it. We saw that global-warming denial or skeptic organizations, generously funded by fossil fuel, corporate, and conservative interests, are actively working to sow doubt about the facts of global warming. These organizations are key players in the fossil fuel industry's strategy to create a "disinformation playbook" designed to confuse the public about global warming and delay any policy action on climate change.

Corporate fossil fuel interests spent huge amounts of money to question the legitimate scientific consensus around climate change and counter established findings by promoting their own pseudoscientific studies that led to self-interested and predetermined outcomes. They paid seemingly independent scientists to conduct this "research" and/or to further undermine legitimate findings in media campaigns and publicity efforts. They, and many of the policy makers who support them, intimidated or openly attacked legitimate climate researchers, relentlessly skewed the analysis of the costs and benefits of any proposed regulations, did everything they could to undermine the legitimacy of the regulatory process itself, and made huge campaign contributions to elect politicians who would do their bidding in the policy process. Why did the fossil fuel industry do these things? To enhance its profits and to protect itself against governmental regulation. The fossil fuel industry saw such tactics as utilitarian means to pursue their practical objectives. It was necessary to provide "alternative facts" and pseudoscience as weapons against science because objective peer-reviewed science would not be of use in the pursuit of its goals. It would be, if believed and taken seriously, an adversary to the interests and goals of the fossil fuel industry. Conflict, a war against climate science in the pursuit of self-interest, was the inevitable choice.

The case of hydraulic fracturing for natural gas, as discussed in chapter 4, followed a different path to the same conclusion. In this case, the science defining the risks and vulnerabilities associated with the new horizontal drilling technologies, and its related activities throughout the process, was in its infancy. We saw that the fracking revolution took off very quickly and very much in advance of serious efforts at risk assessment. Early on, as in the case of Dimock, Pennsylvania, and the contamination of an aquifer that filled household wells in a rural area, it became apparent that more serious efforts at risk assessment were essential in the interest of public health and safety.

Two decades into the fracking revolution, it is sadly true that we have barely begun to do the science that will provide the insight that we and our policy makers need to assess and manage the possible environmental and health risks posed by horizontal fracking technologies and practices. The industry has done its utmost to deny any such threats and discredit any scientific studies that reveal them. The conflict dynamic, as we have seen it played out in the case of fracking, was engaged not to deny an existing scientific consensus but to preempt any future scientific documentation of environmental or health risks. The fossil fuel industry saw this as a necessary political strategy to defend its interests. Any science that might disprove or contradict the industry's assertions that the fracking technology and process were perfectly safe were anticipated and quickly challenged, and very successfully so to date, to prevent government regulation of the industry.

As the conflict surrounding climate change and hydraulic fracturing demonstrate, self-interested corporations and conservative ideologues who oppose all forms of governmental regulation will understandably work to discredit legitimate science where its conclusions support arguments for any policies or regulatory schemes they find objectionable. This discrediting of science includes, as a political and practical necessity, the cultivation of public doubt about the legitimacy of scientific conclusions, which is very effective in slowing the policy process. In a culture where expertise is already regarded with suspicion and resentment, where book learning is deemed elitist and less trustworthy than conspiracy blogs, and in a political climate that has lost the capacity to differentiate between facts and opinions and that regards every fact contrary to opinion as either part of a dastardly plot or "fake news," the cultivation of public doubt about the legitimacy of scientific conclusions is too frequently and too easily accomplished.

As we saw in chapter 4, science has begun to engage some of the concerns surrounding the fracking process. Its findings have become more certain in some areas, and they have demonstrated the need for much more analysis in others. The work of science has been interpreted by the warring parties in the context of their prevailing assumptions and goals. The industry tends to deny risks or be indifferent to them until disaster strikes. Keeping production costs down, avoiding costly or inconvenient regulation, and turning a profit are invariably more important goals. Science is trying to understand and document the risks in order to provide a reliable foundation upon which public- and private-sector decision makers can make sound decisions about the environment, public health, and safety. The conflict dynamic has worked against any genuine cooperation between the two parties. This may be somewhat inevitable, but the prevailing culture and the extreme partisan political environment have combined with this conflict dynamic to create a perfect storm that guarantees the accumulation of environmental and public health disasters that need to be understood, anticipated, and managed. In the present cultural and political climate, public policy is more frequently influenced by massive disinformation campaigns than by science. In fact, the conflict dynamic and the misinformation campaigns succeed too frequently in drowning out scientific knowledge.

The third dynamic we discussed in chapter 5, the resistance dynamic, refers to a situation where strongly held cultural values are in opposition to the findings or conclusions of science. Significant segments of the public may be opposed to science on religious grounds, for example, or for reasons of ideology. As our discussion of evolution and creationism demonstrated, fundamentalist religious belief and science do not always see eye to eye. We saw that there is a complete consensus among scientists around the world that evolution is the backbone of modern biology. It is a historically demonstrable reality as well. Yet, in the name of religious belief, the evidence-based and robust conclusions of science, the facts if you will, are routinely denied or dismissed. Even in the most scientifically advanced nation in the world, the United States, we see the rejection of science in defense of religious belief. This rejection of science or of knowledge is a right we all have, and should always have, one might argue, in a free society. While questioning or doubting can often be a good thing, it is hard to suggest the same about outright rejection or denial of science or about basing public policy on the denial of or resistance to science in the name of belief.

We must also see the resistance dynamic in the context of our culture. We have noted that in our culture, science, the arts, and humanities as the measure for knowledge and quality are increasingly less authoritative than mere opinions. Indeed, we see entertainment, self-righteousness, willful ignorance, and deliberate gullibility are too frequently the foundation for opinions. Worldview or ideology defines reality. Religion is held by many to be superior to knowledge with respect to defining reality. Facts are not regarded as objective realities to be digested, analyzed, and understood but as things to be debated, challenged, and dismissed in favor of beliefs, opinions, and values of a religious or partisan group, a popular leader, or in defense of a self-made reality. Resistance to science is typically born of the wish to avoid or deny a reality science has documented and explained that conflicts with reality as we believe it to be or as we wish it were. On an emotional level, such a reaction may be understandable, but on a rational level, this is not a good foundation for public policy. Yet, as we have seen in state after state even up to the present moment, legislation is routinely proposed to retard science education and to attack science literacy in the name of religion or political partisanship. That this erosion of science in our classrooms is promoted and tolerated in our public discourse, and in some partisan political circles, is yet one more example of how easily scientific knowledge is drowned out in our public dialogue.

One area where we might expect to see more cooperation than conflict, and very little denial, in the relationship between science and politics is in preparing for and responding to public health emergencies. But as our assessment of the panic dynamic in chapter 6 demonstrated, there are a number of factors that work against the relationship. We saw that experts' objective assessments of infectious disease outbreaks and pandemics tell us that the United States and the world need to recognize that we are unprepared to respond to a global medical crisis. We saw that all the informed sources tell us that the United States and the world need to be prepared through enhanced coordination, more investment, and better infrastructure, such as improved systems for disease surveillance. Most policy makers generally understand the importance of all this, but the fact remains that the science and politics of the matter have not interacted smoothly to address the widely perceived and broadly shared need to be prepared. The Ebola epidemic highlighted major inadequacies in the United States' ability to respond to global public health emergencies. Experts concluded that our response to the Zika crisis, as with Ebola, was shaped by the fragmented and partisan U.S. political system, not by epidemiology or medicine.

In our assessment of pandemic preparedness, we saw that the science is incredible in its ability to warn us about public health emergencies. But until such an emergency unfolds, the public is detached or uninterested, and policy makers are very slow to engage. When a pandemic or public health emergency unfolds, the public is in a state of confusion or panic, and policy makers are unprepared and without immediate answers and take a considerable amount of time—too much time, according to the experts—to gear up and respond. A large part of this is explained by the fact that our policy makers and political actors are mostly reactive. They respond to a health crisis, often on a delayed basis or in an inefficient manner, rather than commit the necessary time and resources to anticipating and preparing for one. This almost guarantees that we will not be ready or prepared for a major pandemic or emergency.

Even at its best, the relationship between science and politics is a troubled one. Even where one would expect them to see eye to eye, dealing with a public health emergency for example, they think and operate differently. Scientists and politicians think and communicate differently about everything, and this makes them incompatible when it comes to the identification of priorities, problems, and solutions. It takes considerable effort to get them to address the same reality. When they are collaborating, as during the Cold War and the space race, they are acting for very different reasons, and the interaction is driven less by science and more by practical and utilitarian concerns. Also, the dominant partner will be the political one or the policy maker. In a democracy, one might suggest that this is as it should be. But, sooner or later, as political priorities and governing parties change, science is doomed to become abused or rejected. The panic dynamic is another environment in which science and politics are incentivized to cooperate, and when they sincerely desire to do so, but their habitual patterns of thought and behavior, and the very different worlds in which they live and work, make it impossible. Despite what might be the best of intentions, this leaves the nation and the world less than ideally prepared to combat a global health crisis.

The conflict and resistance dynamics most commonly shape the relationship between science and politics and have the greatest potential for relationship breakdown. Whether science is denied or attacked in the defense of corporate self-interest and partisan advantage or resisted in the name of religious or cultural beliefs, its ability to constructively influence public policy is both contested and limited. Indeed, science will often be completely drowned out by massive distortion and misinformation campaigns. Whatever constructive influence science may have in public policy deliberations, and there will be some even in the most contentious of circumstances, it will be watered down even more by the partisan war raging around it and the jerky forward-and-backward steps taken as one partisan group has a temporary advantage over the other. Within each of the four dynamics we have discussed are embedded the inherent obstacles to an ideal relationship. Within our culture and within our political environment today we see a recipe to combine these inherent obstacles, already embedded in the relationship, into a toxic concoction that will do very great harm.

The inherently troubled relationship between science and politics is made more difficult by our anti-intellectual culture and the hyperpartisanship of our contemporary politics. The cultural and political climates serve only to promote even more intense conflict and resistance to science and to diminish the prospects for any effective collaboration in the public

interest. The cultural and partisan influences make it all the more difficult, impossible perhaps, for science and politics to do anything but collide over the greatest concerns of our age. Science and politics not only live
and work in different worlds but also have come to inhabit two very different and often incompatible realities. This gives rise to what we have previously called the Cool Hand Luke syndrome: "What we've got here is failure
to communicate." For all the complexities in the relationship between science and politics, for all the factors in each of the four dynamics that make
the relationship imperfect and often very troubled, the one thing common
to each dynamic is the lack of a shared reality. This muddies communication and often precludes the effective interaction and cooperation between
the world of science and the world of politics; this has only been com-
pounded by the current cultural and political environment.

The lack of a shared reality is a pressing problem in the United States
today. Liberals and conservatives do not share a reality. Political discourse
today communicates very little beyond a vicious contempt for the "other."
There is no longer any unifying value structure or basic belief that holds
the disparate parts together as one nation. In much the same vein, there is
no fact that is agreed to or accepted. Worse still, there is no common
acceptance of any objective standards for truth. We have a president, and
many partisans, who call legitimate news "fake" and who do not accept
the evidence of science. The lack of a shared reality divides urban from
rural, rich from poor, east from west, north from south, and one race
from another. Everything, including all forms of information and media,
is becoming a weapon in a war of conflicting worldviews. Every discussion on every issue is a standoff in this war of worlds. Climate doubters
clash with proven science. Bathrooms have become battlefields, borders
are battle lines, and walls are preferred to bridges. Sex and race, faith and
ethnicity, and every other conceivable difference is a battlefield as what
we used to call a melting pot is boiling over. In the context of this increasingly toxic environment, the lack of a shared reality between the scientific
and political worlds is an even more serious threat to the future of humanity. At a minimum, a functional and constructive relationship between
science and politics requires that the two worlds must share one reality.
That is made all the more difficult by the cultural and political variables
at play in society at large. But it is the place where the search for solutions
to the Cool Hand Luke syndrome, or the communication problem, must
begin. It is imperative that on some level the two worlds find a way to
bridge the reality gap between them. If they do not, there is no path forward in their relationship. A continuation of their failure to communicate
is simply not acceptable.

Solution: Two Worlds—One Reality

The solution to the Cool Hand Luke syndrome is by no means as easy to address as it might seem. The solution is very complex, and it involves many aspects of our society that influence the relationship between science and politics. The previous chapters have, it is hoped, helped us to understand the importance of the problem to be solved. Reflecting on what has been covered in these chapters, we now discuss what the solution to the problem might ideally entail. What is to follow is but a summary of what, in the context of our present discussion, seem to the author to be the major items that must be addressed. Identifying what must be done and the specifics of getting it done are two very different things. The discussion to follow will focus on the "what must be done" part. The "how to do it" part, as related to each suggestion, is a subject for many books. In fact, the "how" will require a serious and inclusive national dialogue, but that dialogue cannot take place until it is clearly understood what must be done. The main purpose here is to articulate what must be done to *bridge the gap between the scientific and political worlds*. In essence, we will see that there are actually a number of gaps to be bridged.

As we have seen, the very different worlds in which science and politics live and work place many obstacles in the path of communication. These obstacles are greatly complicated by the cultural and political contexts of American life, or so we have argued. One cannot expect to easily change the prevailing tides of the culture or the hyperpartisanship of the present age. Neither can one expect that improving the ability of science and politics to communicate will lead to an ideal relationship between the two. But if doing such might at least enable these two very different and often contentious worlds to acknowledge the same reality, the disagreements and debates that remain to be resolved in the cultural and political arenas might have more meaning. They might be about what really matters. For example, the science supporting the consensus about climate change might be seen as a foundation for a common reality. But this common reality does not resolve all differences of opinion in the policy realm. Debates will be had, need to be had, regarding the different options for responding to this shared or common reality. Debating the options for addressing a reality is both preferable and more productive than debating the scientifically documented reality itself. Communication, which here means bringing the two worlds into one reality, will not end division, partisan disagreement, self-interested argumentation, or conflict. But it would make the remaining debates and discussions far more intelligent and productive. It would ensure that they begin from the same starting point, the same foundation in an objective reality. To do their respective jobs well, scientists and politicians must operate in very different settings. They will always, given their differing perspectives and responsibilities, live in separate worlds. But it is also important they act together to understand, manage, and improve our lives. At a minimum, this does require a shared reality.

Is it possible, one might reasonably ask, to bring science and politics together into a shared reality? It has not quite happened to date, and the evidence of history suggests that it may never happen to the degree that it must. But trying to make it happen has never been more important than it is at this moment in human history. The fact is, we have reached a point when almost every policy decision that will impact the quality of human life will depend on the quality of the relationship between science and politics. Neither scientists nor politicians can be allowed to be blind to that fact. They both must meet a greater responsibility, greater than any previous generation, to bring their worlds together into one reality. They are not alone in this responsibility but central and essential agents in a process that must ultimately involve all of us. So it is with them, scientists and politicians, that we will begin our discussion of the solution.

Scientists have for the most part not seen themselves or the work they do as being connected to the world of political debate over values and policy priorities. Even where they have seen the necessity of their work as a source of valuable information for policy makers, most scientists have limited their sense of personal responsibility to the narrow confines of the scientific work they do. They have assigned the responsibility for the use and application of their work in the policy arena to others. The scientists who worked on the Manhattan Project, for example, did the science to develop a weapon of mass destruction. Perfecting the science to create the bomb was their responsibility. But even among some of those who were most disturbed at the tremendous destructive force this would unleash, the responsibility for its ultimate use was typically assigned to somebody else. The "somebody else" implied here would be the president and the military. This compartmentalization of responsibility is a staple of a culture that typically limits ethical responsibility to a narrowly defined function as opposed to the broader outcomes or impacts that the function contributes to or makes possible. After World War II, J. Robert Oppenheimer, the director of the Manhattan Project, lobbied for international control of nuclear power, advocated efforts to prevent nuclear proliferation, and opposed a nuclear arms race. These things were regarded as outside of the scientific realm. They were political concerns that, as many scientists might typically say, went beyond the appropriate scope of scientific work.

Scientists are apt to say that they see no connection between their work and the political world and the debate over conflicting values in the policy arena. They comfortably limit their focus to the lab and their specific work. They lead enjoyable lives doing interesting and very important work, receive grants and corporate funding, and speak mainly to other scientists. Generally speaking, they do not have to worry too much about or communicate with the rest of the world. Most importantly, as so many of them choose to emphatically maintain, they can choose to stay out of the nasty disputes that inevitably develop among other elements of society, especially in religion and politics. As a result of these attitudes and habits, scientists are a mystery to most Americans. Likewise, the average citizen is a mystery to scientists, whose work environment is separate and isolated from the public. The general public is separated from scientific knowledge and from the scientists cloistered in their labs. One must raise the fundamental question of whether or not this is a good thing. Is it good for scientists? Is it good for average citizens? Is it good for our society?

Many scientists see politics as dirty and participation in public engagement as beneath them. This attitude ascended and seemed to be the norm for a considerable period at the end of the Cold War. Except for the publicizing of results, many in the scientific community basically retreated from public view and became invisible to the average citizen. This meant that very few of the wonders and marvels of science were being fully conveyed to or understood by the public. It meant that as science became more politically relevant and important to meet public policy challenges, it was more misunderstood by the general public than ever. This contributed to the widening gap between the scientific and political worlds that needs to be closed.

It is an uncomfortable truth, at least for most scientists, that science is political. This is to say that the findings of science inevitably *are relevant and connected* to the societal debate about values and public policy. While scientists may wish to refrain from such debate, it is not wise to ignore the fact that the conclusions of science are relevant to that debate. It may provide essential information that must inform the debate. Just as likely, partisans, even those with good intentions, may misrepresent the science to serve their political objectives. We have seen that in our discussions in previous chapters. It is worth remembering that we have a Congress full of lawyers who are trained not to get at the truth but to win an argument. They will do anything necessary to defeat their opponent. They will sacrifice many things, including and especially the truth, to achieve their partisan objectives. Science, we might say, is all about knowledge. Scientific truth, verified through rigorous testing, matters above all else for scientists. Science is unavoidably political because of its relevance to informed discussion and analysis about public policy issues. It is unavoidably political because politics will use or misuse the knowledge it produces to promote partisan agendas. Knowledge is power, and the battle to promote or retard the ability of knowledge to influence public policy is political. While their professional business is knowledge, scientists may need to start paying more attention to politics, not for self-promotion or to protect their parochial interests but to promote the truths that they devote their lives to uncovering.

Like it or not, scientists must do a better job of communicating science to the public and to politicians. Why? Simply put, the gap between the public and science must be bridged if our democracy is to be preserved. Scientific knowledge is extremely important in assessing and addressing most of the important issues in our political system. It should play a major role in addressing public policy challenges. An informed citizenry is a critical necessity in a democracy. The truth is, unfortunately, science is producing knowledge so quickly that the public simply cannot keep up. This knowledge is also complex and difficult for the untrained mind to grasp. Yet, if this knowledge is essential for sound policy, it is important that the public have some access to it so that they can evaluate the political debate and make informed choices in the voting booths across the nation. The volume and the complexity of the scientific knowledge has made it all too easy for self-interested entities to undermine the role of science and data in decision making. Antiscience or propaganda campaigns to discredit science, and to prevent its knowledge from being used most productively in the policy process, have been very successful. Self-governance relies on a well-informed voter, but politics is often about controlling or skewing the information that voters have, even if it means keeping them misinformed or uninformed to advance a specific interest or partisan agenda. If those who have produced the scientific knowledge do not respond to the persistent efforts to undermine this knowledge with propaganda, and if they do not take steps to ensure that the public understands and benefits from the knowledge they have produced, who will? Who could if not they?

This is not to suggest that scientists should become combatants in the political arena or partisan advocates or candidates, although one cannot help but think it would improve the public policy debates and policy outcomes if more scientists did run for and hold public office. It is not even suggested that they become activists in any traditional sense of the word. What is suggested is that scientists help bridge the gap between themselves and the public. They must, in addition to doing the research, do more to ensure that what they learn is *accurately conveyed and understood* by the public and by the people they elect to public office. This means taking an interest in and working to improve the quantity and especially the quality of media coverage of science. It means more effort to engage with the public in order to help them understand not only the things scientists know but how they know them, how the scientific process works, and what its relevance is for policy issues. Closing the gap between science and the public will be the single most important contributor to the ultimate objective of incorporating scientific knowledge into the policy process in the most constructive ways possible. If not that, and on a less idealistic level, it would at the very least diminish the effectiveness of antiscience propaganda campaigns to deny, distort, or misrepresent science.

Scientists often do note that public opinion on scientific issues is too frequently shaped by fear and ignorance about science. As we have seen in some of our discussions throughout this book, a number of surveys do indeed demonstrate a large gap between public opinion and the informed opinion of scientists on a wide range of scientific topics. In recent years, more scientists have actually begun to realize that it's not enough to just do science. Many have come to understand that researchers have to be able to explain their work in words that make the discoveries relevant and understandable to decision makers and to the public. It is scientists' goal to achieve a consensus of rational and evidence-driven opinions about the things they study. A part of their job, it is being argued, must be to find new ways to convey to the world how they reach that rational consensus. Today, and often by design in our cultural and political wars, there is a great confusion in the public perception of how science works and how the conclusions drawn from scientific research may be interpreted. This contributes to the gap between the public and scientists. It really is necessary for scientists to close this gap with knowledge. As has been said, if they don't do it nobody else will, or can, for that matter.

If it can be agreed that our national security, health and well-being, and economic progress are all connected to science and the knowledge that it produces, then those who produce that knowledge must ensure that it is properly understood and applied. With the constant efforts being made in our public discourse to discredit, misrepresent, or deny science for political or material gain, it is necessary that those who produce and understand the science under partisan attack be engaged in repelling this assault in the interest of knowledge and truth. This means that scientists must, no matter how uncomfortable it might make them, actively *promote science in the public discourse*. They must assertively take one fundamental position in the public dialogue. They must be proscience. This proscience approach must always and of necessity be nonpartisan. But nonpartisan is not the same as bipartisan or multipartisan. Being proscience is not neutral either. It does not refuse to take a position. A proscience approach does not care where the partisan divide is or where the work of science falls on it in the context of any given issue. It steadfastly holds to and advocates going with the evidence-based knowledge of science and what it recommends or suggests.

Scientists may recoil from the notion of public engagement as a part of their professional lives. Some will not like the suggestion that they might have a responsibility to explain to the public why their work is valuable and deserves support, but it bears repeating that scientists are the ones who understand the depth and complexity of modern science best. They are in the best position to translate that depth and complexity to nonscientists. That may sound like a burdensome obligation, but it is also a necessity for securing the future of scientific progress. It is wise to remember that it is the public and its elected representatives who ultimately determine the fate of funding for basic scientific research. It is perhaps understandable that scientists may be reluctant to engage the public if one considers that the public is as alien to them as they are to the public. In other words, just as the public must be educated on scientific topics, the scientific community may need to become better educated on the things that shape public attitudes and opinions.

To improve their interactions with the public, scientists must understand how the public thinks, how its opinions are formed, and the variables that influence the formation of opinions. Those who care about and wish to improve the role of scientific knowledge in public policy must recognize the need to work at understanding the shaping of public sentiment. Shaping public sentiment is something of a science that politicians and special interests have honed into an art form. To win any contest with science deniers and propagandists, one must know how their work is done and how to successfully counteract it. Charts, data, and scientific jargon are not things that will get this job done. The things scientists have learned, the things they know, are complex. This complexity is very often beyond the reach of most nonscientists. The complexity of scientists' knowledge, and the scientific language they use to explain it to each other, is powerless to communicate with the general public. This is not to say that the general public is stupid. It is simply to say that the people who comprise the general public are not, by and large, scientists. The challenge will be to convey what science knows in a manner that is approachable and nonintimidating to nonscientists.

If scientists wish to convey the knowledge and the methods of their craft to the public, and it is here regarded as a necessity, they might benefit from some social scientific understanding of public opinion. They might come to see the value of using data from social scientists to better understand public attitudes toward science and technology. They might recognize the need for some training in mass communications so that they can communicate as effectively as the media-trained partisan hacks and corporate shills, and they will have to design better messaging for the public. Scientists will need to learn about the values of the audiences they must target in their efforts to shape public sentiment. They will also need to learn how to build public trust and how to listen to and respond to public concerns.

The closing of the gap between scientists and the public is but one important step in the process of creating a single or shared reality between science and politics. Why? To the extent that science can effectively communicate with and educate the public and politicians, it will significantly diminish the prospects for successful disinformation or science denial campaigns. It will create more public capacity to place reasoned and scientifically grounded demands on policy makers. An informed citizenry will provide more incentives for politicians to eschew science denial or scientific disinformation campaigns in addressing public challenges and policy options. But much more is required to complete the task of bringing two worlds into one reality. Politicians also have a fundamental role to play and much work to do. Like the general public, politicians will need some help from the scientific community.

Politicians have varied educational backgrounds, and their scientific literacy is typically minimal. In fact, it is all too frequently laughably minimal. Politicians, we have repeatedly said, deal with the minutia of the moment. They want quick, tangible, and even spectacular results. They would like research to be connected to a useful application that advances their ideological view or partisan advantage. To this broad overview must be added the willingness of many candidates and office holders to abandon or deny science outright to support partisan agendas or vested corporate interests. As we saw in our discussions of evolution, climate change, and other issues, antiscience has been a part of our political and policy dialogue forever. Sadly, especially in relation to so many of our contemporary challenges, we have seen that antiscience or science denial influences public policy as much, sometimes more, than does scientific knowledge. Too often knowledge is not the basis for public policy. Too often the basis for policy choices is determined by a political agenda that seeks only the facts that support it and is not going to entertain any evidence or data that does not support it. This makes scientific or knowledge-based decision making much more difficult. This is the curse of the antiscientific revolution that seems to forever be bursting out at the most inconvenient times in our most important policy debates. When public policy debates become a clash between science literacy and science illiteracy, politicians are rarely held accountable for it. This is a problem that must be addressed if the relationship between science and politics is to be improved.

Politicians often ask scientists for their expert opinion. Sometimes they even try to make informed judgments based on those opinions as they formulate public policy initiatives. Even in instances where policy makers may have fundamental disagreements, we would like to believe that science could serve to simply inform the debates that are of the greatest consequence to us. But it doesn't always work that way. Rather than informing the debate, science is too frequently politicized and used as a tool in political debate. Partisans on both sides bring in their own "experts" and their own "expert analysis" of data, not to get at the scientific truth of any matter so much as to further tactical political goals. This results in the development and presentation of obviously skewed "science" that tailors reality to conform to a desired partisan or ideological outcome. It is, of course, easy to blame politicians for this state of affairs, but tactical political maneuvering is their job. Winning the argument, as we have said, is the objective of politics. It is difficult, in most political circumstances, for science to be anything other than a tool of the partisan debate. This does not always result in the sin of science denial—as it does in the climate debate, which trots out the same handful of anticlimate change or pseudoscientific "research" at every congressional hearing— but it does great damage to the ability of science to be of real use in guiding public policy.

What should the relationship between science and public policy be? Most people, including scientists, would say that agreeing on scientific knowledge is a prerequisite for reaching political consensus and taking intelligent policy action. Ideally, science reduces uncertainties and increases our understanding. As our understanding increases, we are ideally enabled to adjust our assumptions about problems and how they should be addressed. This is an attractive linear argument, but the policy process simply does not work this way. The ideal seems to suggest that scientific advice to policy makers can be restricted to technical issues and that subjective values are irrelevant to decision making. It is naive, against the backdrop of politics and the partisan practice of cherry picking or tailoring science to conform to a predetermined reality, to think that scientific advice results in any sort of political consensus. Facts, scientific or any other, are massaged and made into weapons in the contest to win partisan advantage. So we have a dilemma. There is no doubt that scientific thinking, based on logic and evidence, could improve the decisions policy makers make on most issues. But especially in today's political and cultural climate in which all facts are contested and there is no objectively agreed upon foundation for reality, scientific thinking is not always a strong influence on policy makers.

Scientific evidence, of course, is not the only consideration in a policy decision. For complex issues, many interests have to be balanced. There are also many situations where the science itself is uncertain. But where science is relevant as a component in the policy process, and where the science is reasonably settled and clear, it still has difficulty being heard, and its influence is weakened in the partisan skirmishing to define reality to fit a predetermined partisan outcome. There are no easy and quick fixes that can bring science and politics into the same reality, but there are steps that can begin to move them a bit closer to that objective. Just as the gap between scientists and the public must be closed, the gap between scientists and politicians must be bridged.

As a first step, science must be more than an afterthought or a nuisance to be avoided in political debates, election campaigns, and politician-constituency dialogues. Given the importance of science to our economic future, our national security, and our basic well-being, many have already suggested that candidates for public office should be required to engage in science debates. This makes a great deal of sense. Scientists in particular should promote this requirement, or perhaps "expectation" is the better word for those who think a "requirement" is a bit too much. But the point is that these candidates, when elected, are going to be making decisions about climate change, energy, the environment, health care, stem-cell research, biosecurity, nuclear weapons, science education, the teaching of evolution, and many other issues where science is relevant and important. These science debates need not require that candidates respond as a scientific expert might, but they should provide a measure of a candidate's general awareness and assessment of major issues from a relevant scientific and evidence-based perspective. These debates should generally demonstrate how well they understand the scientific method and how they assess its relevance to the decisions they must make if elected. Disinformation or uneducated opinion, not to mention partisan inanities, may sneak into science debates just as they do in political conversation generally. But these things can be eliminated to some extent if the questions and topics in the debate are designed to incentivize candidates to base their responses on knowledge and to disallow pandering to antiscience political audiences. This might be accomplished, some proponents of science debates have suggested, by having a panel of scientists develop the questions and science communicators serve as moderators.

The concept of science debates has some appeal to scientists and some citizens perhaps, but journalists, editors, news directors, and networks will no doubt think it unappealing or of little interest to the public. Politicians themselves would no doubt seek to avoid such debates because they would regard them as too risky and as having little benefit in promoting the causes of their electoral base or the goals of their partisan agenda. Science has evolved over many centuries to become an integral part of modern society, underpinning our health, wealth, and security, yet scientific evidence is often willfully disregarded by politicians in the United States and worldwide. This is the specific thing to hold politicians accountable for when all is said and done. It is important, now more than ever, to reinforce with politicians the need to value and respect science in the development of evidence-based policy. Politicians represent and reflect their constituencies and their partisan preferences. That is appropriate in a free and democratic society. But politicians also must come to see that they have the added responsibility of making informed decisions and public policy based on sound evidence. Politicians cannot simply reflect the tendency of people to allow their worldview to influence which scientific facts they believe. They must be seen to have, and they must see themselves as having, a greater responsibility to ensure the ongoing contribution of science to government decision making and thereby to enhance the role of science in our society. If, however, they are unwilling to participate in science debates or discussions, what other means do we have to hold them accountable for this greater responsibility?

It has already been said that scientists must work to close the gap between themselves and the public. They must also seek to close the gap between themselves and politicians. Individual scientists need to play a larger role in everyday life to communicate their science, whether to key decision makers or the wider community, in order to counter the "alternative facts" that distract both the public and the policy maker from important scientific truths. Scientists with informed perspectives of the state of scientific knowledge on climate change, genetically modified foods, nuclear science, evolution, and every other area where scientific knowledge is needed to inform policy, can contribute to understanding in the wider community. In doing this, they arm the wider community with enough basic knowledge so that it can hold politicians accountable at the polls. The news media has a role to play here as well. It too must hold politicians accountable for improving the contribution of science to government decision making.

When a politician deliberately misrepresents scientific research simply to ridicule it for political gain, it raises several important questions: Why would a supposedly intelligent person, whom citizens entrust to represent their interests, trash scientific research in this way? If this trashing of science is not generally and accurately reflective of scientific opinion shared among intelligent people in all walks of life, then is it a misrepresentation and an abuse of the public trust? Should politicians be held accountable for this abuse? It is the argument herein that they should be held accountable, of course. But holding them accountable requires that the relationship between science and politics be front and center in the public discourse. Science debates during election campaigns, scientists contributing to enhanced understanding in the wider community, and media vigilance holding politicians to account for the relationship between science and public policy are all essential to the goal of incentivizing politicians to meet the greater responsibility of improving the contribution of science to government decision making. These things will make quality scientific advice such a valuable commodity for policy makers that more and more of them will seek out and hire expert science advisors on their staffs. These things will encourage those with appointment powers or responsibilities with respect to scientific agencies and departments to appoint and/or approve appointees with legitimate and strong scientific credentials.

Politicians will meet what we have called their greater responsibility to ensure the ongoing contribution of science to government decision making only if incentivized (or forced?) to do so. We have talked about the role of scientists and an informed public in helping politicians to meet the greater responsibility. In this context, we must also discuss the role of the news media. Here, too, some changes will be necessary if the news media will be of any use in helping or incentivizing politicians to meet the greater responsibility. Many journalists are apt to say that it is not their job to establish the truth; it is their job merely to relay information fairly. This is a hands-off perspective that suggests that it is not their job to call out politicians or vested interests who circumvent the truth and present "alternative facts" or false evidence to achieve a desired outcome. Even where they may seek to separate fact from fiction, the predominant mode of the news media is to look for a conflict angle. There are two sides to every story, they might say, and presenting both sides is the fair and balanced way to present the news. An interesting and important story is thought to be one where controversy rages, headlines are generated, web clicks zoom, more newspapers are sold, and talking heads can spend hours attracting an audience to hang on every word of a sensationalized narrative that resembles the color commentary at an exciting sporting event. This approach is simply inept when it is applied to scientific questions.

When it comes to many scientific issues, the journalistic approach tends to skew public policy discussions in counterfactual directions. The case of climate change is an example. The overwhelming consensus (97 to 99 percent of climate scientists) says anthropogenic climate change is a reality. Yet, at least until very recently, the major media have not accurately reflected that fact. They have frequently presented, in the interest of being fair and balanced, equal time to those who deny the scientific consensus. In almost every cable news conversation, the pairing of a climate change denier with a climate scientist was such a regular occurrence that the public was enticed to believe that the science was not settled. Often, in being fair and balanced, journalism does not attempt to establish the reality or the truth of a story or a point of view. The result is too often a false "balance," presenting both sides as though they are equal with respect to the weight of the evidence. Objectively speaking, as with climate change, this sort of "balance" leads to an impression that is simply not true. Often, there really aren't two sides to a story. This is especially the case where science is concerned. The earth really does travel around the sun. It completes a solar orbit every 365.256 days. Apparently, "Flat Earthers" are getting some press these days, but the earth is not flat!

The relationship between science and the media has also been characterized as a gap that needs to be closed. Like the gap between scientists and the public, and scientists and politicians, it must be addressed. Often, in their interaction with the media, scientists find themselves in a world ultimately structured by journalistic mass media. They find that they have to adjust to the logic of the media to attract attention. To some extent, that is unavoidable, and scientists must learn how to communicate in that world. But at some point, if it is to have any hope of doing its job fairly and adequately, journalists and the media must adjust to the scientific method and report its findings accurately. At the very least, journalists must give as much attention to convey the reality or the truth of science as they do to presenting "both sides." Where the weight of the evidence is indisputable, neither political partisan nor public office holder should be receiving "balanced" treatment. This balanced coverage is irresponsible where the partisans or the officeholders are demonstrably and objectively wrong. This prioritizing of balanced treatment in the interest of fairness is fine with respect to reporting on politics, editorial commentary, and numerous other stories, but when reporting on events or issues that are defined by or connected to scientific study, it is the absolutely wrong priority.

To be fair, the media has never seen its assignment as educating the public about science or even covering science. Journalists typically monitor science for developments that may be of interest to their readers or viewers. Even science reporters, and these are actually fewer and fewer in number, will often forego stories that are about scientifically important discoveries in favor of trivial stories that will be of greater general interest to readers and viewers. The more significant or important developments are often ignored because the public is assumed to be uninterested. Reader or viewer supposed interests dictate what is covered and reported, not science. In other words, the media in its quest for readers or viewers assumes that it cannot afford to cover what scientists consider to be important. The potential of a story with respect to the number of readers, viewers, and clicks usually determines what they cover. That is a practical and utilitarian assessment in the context of their business. This places even more of a burden on scientists who wish to communicate to the public through the media, and, as previously stated, scientists will have to learn how to play the media game. But as difficult as it will be to change their game, the media also must bear some responsibility for improving its coverage of science.

It is not unreasonable—assuming that science really is an important component in policy making as it relates to our health, economic advancement, and security—to expect that the media might want to improve the quality of their science reporting by hiring more journalists or reporters with science backgrounds, assigning more journalists or reporters to actually cover science, including actual scientists in interviews and on panel discussions, and at the very least making an effort to eliminate the problem of a false balance in science reporting by more clearly separating fact from fiction. This does not mean that scientific disagreements, uncertainties, or debates should be ignored, downplayed, or not covered accurately and completely, but it does mean that science denial and partisan distortions of science need to be identified and called out. This has gotten harder to do with budget cuts at major newspapers and other media. There are fewer and fewer science journalists working for traditional print and broadcast media, and that is a trend that needs to be reversed.

The purpose of science reporting must be to render very detailed, specific, and often jargon-laden information produced by scientists into a form that nonscientists can understand and appreciate while still communicating the information accurately. It is difficult to achieve this purpose without skillful and trained science reporters. But there is never any guarantee that the media will do an adequate job in achieving this purpose. We have all undoubtedly seen examples where science reporting has been pointless, simplistic, boring, or just plain wrong. Sometimes a scientist whose research has become the subject of a news story can be heard to complain, usually with some anger, that the headline and the lead were grossly misleading if not sensational. This speaks to the need, perhaps, for all journalists to be scientifically literate. For example, the inability to distinguish between correlation and causation could easily lead to some very inaccurate reporting. These sorts of problems will always be a fact of life in the relationship between science and the media, but that does not excuse journalists or scientists from working to improve the quality of science reporting. Despite any imperfections that will inevitably remain, journalists, the media generally, and scientists must work to improve the coverage and reporting of science to the public. This is essential for understanding the public policy challenges of our time. It is essential if we are to have any hope of holding policy makers accountable. It is essential if we hope to stop skewing public policy discussions in counterfactual directions.

Perhaps the best way to close the gaps between science and politics, between science and the media, and between science and the public is to do a better job of promoting science literacy. Science literacy is critical to addressing the challenges we face as we increasingly turn to the application of new knowledge and technology for solutions to current and emerging health, social, and environmental challenges. It is, of course, the human capacity to develop and use technologies that has generated many of the contemporary challenges we face. To not work hard to improve our collective societal understanding of science and technology, it would logically follow, is to limit our potential to address the serious challenges we face in a meaningful way. We see with alarming frequency that things like ideology, partisan identification, distortions on social media, and fear can shape our opinions about complex matters. This should reinforce our determination to promote basic scientific literacy. A populous that is ignorant of how evidence is produced and how to critically appraise rhetoric and hype or to distinguish fact from fiction is at risk of making suboptimal choices. All this speaks to the need to do a better job in the American education system of producing science literacy.

To the extent that our schools may not have a genuine commitment to the requirement of science literacy in the curriculum, they do a great disservice to their students and to their communities. This is not to suggest that all students must focus exclusively on STEM subjects (science, technology, engineering, mathematics) in their tertiary education and career choices. It is simply to say that *all students need basic science literacy*. They will need it in their working lives in most cases, and they will surely need it if they are to have any hope of being informed and capable citizens.

Improving education in math and science is only partially about producing the engineers, researchers, scientists, and innovators who are going to help transform our economy and our lives for the better. It is about much more than that. It is, most importantly in the context of our discussion throughout this book, about producing an informed citizenry in an era when many of the problems we face as a nation are, at their root, scientific problems.

As noted in chapter 6, we see some erosion in our commitment to science education when we have politicians and citizens advocating for the teaching of religion (i.e., creationism, intelligent design) in science classes. We see erosion also in the form of lowered standards and inadequate curricula that allow Americans to graduate from high school and even from college with a pathetically weak understanding of science, our history, our governmental institutions, and our economy. In general, according to many studies, education in this country is falling short of what is required to keep the United States productive, stable, and secure. The ability to bridge any gaps between science and politics requires that we see it as failure if the United States does not enroll all students in and enhance their performance in core mathematics and science curricula. Such a failure would only enhance the decline of skill levels in mathematics and science required for postsecondary education, future employment, and informed citizenship.

As should be obvious as we come to the end of our discussion about the solution, if we are serious about bringing science and politics into the same reality, we must strive to close the gaps between four fractured elements in contemporary American life. Scientists, politicians, the media, and the public all have important roles to play if we are to find our way toward knowledge and away from distortion and ignorance. So much of our current cultural and political environment works against this. It will take considerable effort, and some might say it will be impossible. Each of the things we have just discussed, each gap that must be closed, will require a societal effort that many feel is beyond our capacity. Even if it is within our capacity to bridge these gaps, the details will be much more challenging than even our discussion here has been able to convey. But assuming we recognize what must be done, the prospects for a meaningful discussion about how to get it done will improve. With that improvement, prospects for success will be enhanced.

Conclusion

We began in chapter 1 by saying it should be possible for natural science to provide informed opinions about the plausible consequences of our actions (or inactions) and monitor the effects of our choices. We said it should be possible for policy makers to avoid making ill-informed decisions even as they represent different and competing interests. We said it should be possible to respect different values and yet factor the best scientific information into our thinking about public issues. We said it should be possible to improve the communication of science to policy makers and to the public. We said it should also be possible for scientists to better understand politicians and the public. But as our discussion has tried to demonstrate, for all these good things that are possible, we have some very difficult but necessary work to do.

First, we must understand that the dynamics that animate the relationship between science and politics in our society are not typically conducive to creating an ideal relationship. At its core, politics is about getting a group of people to agree on a common course of action. But if you think about it, there is no wholly rational and evidence-based source of truth or legitimacy in any political system. All political systems, to some degree, resort to dogma in some form. Dogma is defined as a belief that is beyond question. Religions are dogmatic by nature and thus are frequently co-opted as a source of political legitimacy. Ideologies sometimes approach religious levels of dogmatic content and can also be used to legitimate authority. Science, on the other hand, is a method for seeking and establishing the evidence-based truth about the physical universe. This often means asking the sorts of questions that dogmatic thinking may find unwelcome and even threatening. Worse still, if science produces knowledge that disagrees with dogmatic assertions, it will meet resistance or stimulate conflict. This is a natural tension that exists between science and politics. Politics (which is based on dogma) and science (which begins with questions) are always going to come into conflict. Inevitably, politics is going to suppress science where its conclusions disagree with its dogma. The political practice of suppressing science exists even in liberal democracies, including and most notably in the United States. Science is strong only when dogma is weak. To the extent that political arguments rely on dogma as a source of legitimacy, they will inevitably conflict with science and scientists. This is when political actors are most inclined to suppress science.

As we examined the four major dynamics, the motivating or driving forces in the policy process that shaped the relationship between science and politics, we saw many examples of the natural tension that exists between science and politics. We saw that the politics, or the dogma, often weighed more heavily than the science in defining the relationship between the two. In the collaborative dynamic, we saw that the politically defined objective (shaped by Cold War dogma and experience) led to a marriage of convenience and a partnership of sorts between politics and science. In the conflict dynamic, we saw that both ideological beliefs and material or economic interests contributed to corporate and political efforts to suppress science. In the case of both climate change and hydraulic fracturing, we saw that science was not the most influential voice in the policy dialogue. Often, it was seen as a threat to political or economic interests and not a partner to be embraced. In the resistance dynamic, we saw the tension between science and religion, and the rejection of science by the most dogmatically religious, play itself out in the debate about evolution. Even in the panic dynamic, a time when scientists and politicians may be most inclined to work together, we saw that the different worlds in which they live and operate remained in conflict.

What we have called the natural tension between science and politics, the product of the very different realities of scientific and political life, is a problem made much worse by the cultural and political excesses we have discussed in this chapter. In a nation gripped by a widening partisan divide and where ideological and dogmatic thinking is on steroids, the divide between the two worlds of science and politics is growing dangerously fast at precisely that moment when it is essential that we work harder to close it. Just when scientific facts are most needed, we see in the United States a growing disdain for scientific expertise. Indeed, the popular media is presently filled with stories suggesting that the U.S. government under the leadership of a new president has launched a blitzkrieg against science and the earth's climate.[18] This may strike many as overstated and emotional language, but it is not an overstatement to say that priority one must be to bring science and politics together into a shared reality.

Science and politics will always be different worlds, so to speak, but reality should to the greatest extent possible be the same for both. When the two cannot agree on the evidence-based foundation for what is real or true in or about the world, it is impossible for science to inform policy makers or for policy makers to mediate intelligently between the two worlds to the public's best advantage. The recommendations we have made in this chapter are only a beginning. They seem simple, but they will be difficult to implement. Scientists must work to close the gap between themselves and the public and between themselves and politicians. Politicians must make wiser use of what science can provide for them as they analyze policy options. The media must be a part of the solution by more accurately and intelligently reporting about science. As a society, we all must demand a more scientifically literate public and support the initiatives that can achieve that goal. But, and this cannot be emphasized enough, these steps are just a beginning. They are necessary to equip all of us—politicians and the public, scientists and nonscientists—to make the necessary distinction between opinion and fact. Only then will we be equipped to render informed judgment and make wise decisions about the most important issues confronting humanity.

There are really no easy answers to the problem in the relationship between science and politics, but recognizing the existence of this problem and understanding its gravity is an important thing. Taking steps to address the problem, or refusing to address it, is where our fate as a democratic society and our future as a species will be determined. When science and politics collide, we all lose. If we can enable them to inhabit the same reality, we may be able to start winning again.

Notes

Chapter One

1. Mason, S. and Weingart, P., eds. (2005). *Democratization of Expertise? Exploring Novel Forms of Scientific Advice in Political Decision Making.* New York: Springer Business and Media.

2. Silver, H.J. (2005). "Science and Politics: The Uneasy Relationship." *Open Spaces Quarterly* 8:1. http://open-spaces.com/article-v8n1-silver.pdf (accessed July 25, 2016).

3. Ibid.

4. "A Failed Experiment." *The Economist,* September 9, 2013. http://www.economist.com/blogs/democracyinamerica/2013/09/science-and-politics (accessed July 25, 2016).

5. Ibid.

6. Silver, "Science and Politics: The Uneasy Relationship."

7. Ibid.

8. "A Failed Experiment."

9. Jacques, P.J., Dunlap, R.E., and Freeman, M. (2008). "The Organization of Denial: Conservative Think Tanks and Environmental Skepticism." *Environmental Politics* 17:3, 349–385.

10. Cook, J. and Lewandowsky, S. (2016). "Rational Irrationality: Modeling Climate Change Belief Polarization Using Bayesian Networks." *Topics in Cognitive Science* 8:1, 160–179.

11. Ibid.

12. Bella, R.N., et al. (2007). *Habits of Heart: Individualism and Commitment in American Life.* Berkeley, CA: University of California Press.

13. Gans, H.J. (1988). *Middle American Individualism: Popular Participation and Liberal Democracy.* New York: Oxford University Press.

14. Ibid.

15. Truman, D. (1951). *The Governmental Process.* New York: Cornell University Press.

16. Schlozman, K.L., Verba, S., and Brady, H.E. (2012). *The Un-heavenly Chorus: Unequal Voice and the Broken Promise of American Democracy.* Princeton, NJ: Princeton University Press.

17. Maasen, S. and Weingart, P. (2005). "What's New in Scientific Advice to Politics?" in *Democratization of Expertise? Exploring Novel Forms of Scientific Advice in Political Decision Making*, pp. 1–19. (Maasen, S. and Weingart, P., eds.). New York: Springer Business and Media.

18. Ibid.

19. Keller, A.C. (2009). *Science in Environmental Policy: The Politics of Objective Advice.* Cambridge, MA: MIT Press.

20. Bijker, I.E., Bal, R., and Hendriks, R. (2009). *The Paradox of Scientific Authority: The Role of Scientific Advice in Democracies.* Cambridge, MA: MIT Press.

21. Brown, M.B. (2009). *Expertise, Institutions, and Representation.* Cambridge, MA: MIT Press.

22. "A Failed Experiment."

Chapter Two

1. Johnson, L.B.J. (1971). *The Vantage Point: Perspectives of the Presidency, 1963-1969.* New York: Holt, Rinehart, and Winston; Gurney, G. (1975). *The Launching of Sputnik: The Space Age Begins.* New York: F. Watts.

2. Gurney, *The Launching of Sputnik*; Halberstam, D. (1993). *The Fifties.* New York: Fawcett Columbine.

3. Gurney, *The Launching of Sputnik*.

4. Ibid.

5. Erickson, M. (2005). *Into the Unknown Together: The DOD, NASA, and Early Spaceflight.* Maxwell Air Force Base, AL: Air University Press.

6. Goodpaster, A.J. (1957). "Memorandum of Conference with the President." October 9. https://www.eisenhower.archives.gov/research/online_documents /sputnik/10_16_57.pdf

7. Ibid.

8. Erickson, *Into the Unknown Together*.

9. Halberstam, *The Fifties*; Erickson, *Into the Unknown Together*.

10. Halberstam, *The Fifties*.

11. Gurney, *The Launching of Sputnik*.

12. Ibid.

13. Erickson, *Into the Unknown Together*; Howell, E. (2012). "Explorer 1: The First U.S. Satellite." Space.com. http://space.com/17825-explorer-1.html (accessed August 9, 2016).

14. Gurney, *The Launching of Sputnik*; Halberstam, *The Fifties*.

15. Halberstam, *The Fifties*.

16. Gurney, *The Launching of Sputnik*.

17. Halberstam, *The Fifties*.

18. Ibid.; Erickson, *Into the Unknown Together*.

19. Kearns, D. (1976). *Lyndon Johnson and the American Dream*. New York: Harper & Row.

20. Ibid.

21. Ibid.

22. Gurney, *The Launching of Sputnik*; Halberstam, *The Fifties*; Erickson, *Into the Unknown Together*.

23. Kaufman, B. (2004). "Lunar Skeptics." *The American Enterprise*, December.

24. Gurney, *The Launching of Sputnik*; Kaufman, "Lunar Skeptics."

25. Neufeld, M.J. (2007). *Von Braun: Dreamer of Space, Engineer of War*. New York: Knopf; Piszkiewicz, D. (1998). *Wernher von Braun: The Man Who Sold the Moon*. Santa Barbara, CA: Praeger.

26. Neufeld, *Von Braun: Dreamer of Space, Engineer of War*.

27. Ibid.

28. Piszkiewicz, *Wernher von Braun: The Man Who Sold the Moon*.

29. Rhodes, R. (1996). *Dark Sun: The Making of the Hydrogen Bomb*. New York: Simon & Schuster.

30. Reeves, R. (1993). *President Kennedy: A Profile in Power*. New York: Simon & Schuster.

31. Ibid.

32. Ibid.

33. Ibid.

34. Ibid.

35. Ibid.

36. Ibid.

37. Ibid.

38. Ibid.

39. Ibid.

40. U.S. News and World Report. (1969). *U.S. on the Moon: What It Means to Us*. Washington, DC: U.S. News and World Report.

41. Ibid.

42. Reynolds, D.W. (2013). *Apollo: The Epic Journey to the Moon, 1963-1972*. Minneapolis, MN: Zenith Press.

43. Shepard, A. and Slayton, D. (1994). *Moon Shot: The Inside Story of America's Race to the Moon*. Atlanta, GA: Turner Publishing.

44. Teitel, A.S. (2012). "Apollo 1: The Fire That Shocked NASA." *Scientific American*, January 2. http://blogs.scientificamerican.com/guest-blog/apollo-1-the-fire-that-shocked-nasa/ (accessed September 19, 2016).

45. Ibid.

46. Ibid.

47. Portree, D.S.F. (2012). "Apollo Science and Sites (1963)." *Science*, April 20. www.wired.com/2012/04/apollo-science-sites-1963/ (accessed August 16, 2016).

48. Ibid.

49. Ibid.

50. Ibid.

51. Ibid.

52. Newton, D.E. (2014). *Science and Political Controversy: A Reference Handbook*. Santa Barbara, CA: Praeger.

53. Reynolds, *Apollo: The Epic Journey to the Moon, 1963-1972*.

54. Newton, *Science and Political Controversy: A Reference Handbook*.

55. Ibid.

56. Ibid.

57. Wilson, J.R. (2008). "Space Program Benefits: NASA's Positive Impact on Society. NASA: Fifty Years of Exploration and Discovery." National Aeronautics and Space Administration. http://www.nasa.gov/50th/50th_magazine/benefits .html (accessed September 14, 2016).

58. Ibid.

59. Ibid.

60. Liptak, A. (2015). "The Real Story of Apollo 17 . . . And Why We Never Went Back to the Moon." http://io9.gizmodo.com/the-real-story-of-apollo-17-and -why-we-never-went-ba-1670503448 (accessed September 14, 2016).

61. Ibid.

62. Ibid.

63. Ibid.

64. Ibid.

65. Wormald, B. (2014). "Americans Keen on Space Exploration, Less So on Paying for It." Pew Research Center. http://www.pewresearch.org/fact-tank/2014 /04/23/americans-keen-on-space-exploration-less-so-on-paying-for-it/ (accessed September 28, 2016).

66. Ibid.

67. Ibid.

68. Ibid.

Chapter Three

1. Sullivan, W. (1962). "Books of the Times." *New York Times*, September 27, p. 35.

2. "The Story of Silent Spring." National Resources Defense Council, August 13, 2015. https://www.nrdc.org/stories/story-silent-spring (accessed October 31, 2016).

3. Ibid.

4. Thorpe, W.H. (1963). "Where Every Prospect's Vile." *The Observer*, February 17.

5. Thoreau, H.D. (1854). *Walden; or, Life in the Woods*. Boston, MA: Ticknor & Fields.

6. Stanko, N. (2011). "Environmental Movement." Greeniacs.com, April 22. http://www.greeniacs.com/GreeniacsArticles/Environmental-News/Environmental -Movement.html (accessed November 7, 2016).

7. Ibid.

8. Shabecoff, P. (1988). "Global Warming Has Begun, Expert Tells Senate." *New York Times*, June 24. http://www.nytimes.com/1988/06/24/us/global-warming-has-begun-expert-tells-senate.html (accessed November 16, 2016).

9. Ibid.

10. Christianson, G. (2000). *Greenhouse: The 200 Year Story of Global Warming.* New York: Penguin Books.

11. Dressler, A. and Parson, E.A. (2010). *The Science and the Politics of Global Climate Change.* Cambridge, UK: Cambridge University Press.

12. Stone, B. (2012). *The City and the Coming Climate.* Cambridge, UK: Cambridge University Press.

13. Arrhenius, S. (1896). "On the Influence of Carbonic Acid in the Air upon the Temperature of the Ground." *Philosophical Magazine and Journal of Science* 41, 237–276.

14. Lallanilla, M. (2013). "What Is the Keeling Curve?" *Live Science*, May 2. http://www.livescience.com/29271-what-is-the-keeling-curve-carbon-dioxide.html (accessed November 16, 2016).

15. Ibid.

16. Stone, *The City and the Coming Climate.*

17. Ibid.

18. Ibid.

19. Peterson, T.C., et al. (2009). "State of the Climate in 2008." *Bulletin of the American Meteorological Society* 90:8, S17–S18.

20. Allison, I., et al. (2009). "The Copenhagen Diagnosis: Updating the World on the Latest Climate Science." Sydney, Australia: UNSW Climate Change Research Center, p. 11.

21. Kwok, R. and Rothrock, D.A. (2009). "Decline in Arctic Sea Ice Thickness from Submarine and ICESAT Records: 1958-2008." *Geophysical Research Letters* 36, paper no. L15501.

22. National Snow and Ice Data Center World Glacier Monitoring Service. http://nsidc.org/sotc/sea_ice.html (accessed July 9, 2013).

23. NOAA Weather and Climate Extremes, https://www.gfdl.noaa.gov/extremes/ (accessed November 22, 2017).

24. Kreft, S., Eckstein, D., and Melchior, I. (2017). "Who Suffers Most from Extreme Weather Events? Weather-Related Loss Events in 2015 and 1996 to 2015." Germanwatch. https://germanwatch.org/de/download/16411.pdf (accessed January 11, 2017).

25. United Nations Environmental Programme. (2016). "The Adaptation Gap Report 2016." http://web.unep.org/adaptationgapreport/2016 (accessed January 11, 2017).

26. NASA. (2017). "The Consequences of Climate Change." Vital Signs of the Planet. http://climate.nasa.gov/effects/ (accessed January 12, 2017).

27. Ibid.

28. Stranahan, S. (2008). "Melting Arctic Ocean Raises Threat of Methane Time Bomb." YaleEnvironment360. http://e360.yale.edu/features/melting_arctic_ocean_raises_threat_of_methane_time_bomb (accessed February 3, 2017).

29. Whiteman, G., et al. (2013). "Climate Science: Vast Costs of Arctic Change." *Nature* 499, 401–403.

30. Ibid.

31. Eilperin, J. (2016). "Trump Says Nobody Really Knows if Climate Change Is Real." *Washington Post*, December 11. https://www.washingtonpost.com/news /energy-environment/wp/2016/12/11/trump-says-nobody-really-knows-if-climate -change-is-real/ (accessed February 6, 2017).

32. Mufson, S. and Eilperin, J. (2016). "Trump Transition Team for Energy Department Seeks Names of Employees Involved in Climate Meetings." *Washington Post*, December 9. https://www.washingtonpost.com/news/energy-environment /wp/2016/12/09/trump-transition-team-for-energy-department-seeks-names-of -employees-involved-in-climate-meetings/ (accessed February 7, 2017).

33. Schlanger, Z. (2017). "Rogue Scientists Race to Save Climate Data from Trump." *Rincon Tech News*. https://longislandtechnologynews.com/2017/01/rogue -scientists-race-to-save-climate-data-from-trump/ (accessed November 22, 2017).

34. Ibid.

35. The White House. https://www.whitehouse.gov/america-first-energy (accessed January 20, 2017).

36. Revelle, R., et al. (1965). "Atmospheric Carbon Dioxide." President's Science Advisory Committee, Panel on Environmental Pollution, Restoring the Quality of Our Environment: Report of the Panel on Environmental Pollution. Washington, DC: The White House.

37. "Special Message to Congress on Conservation and Restoration of Natural Beauty." American Presidency Project, February 8, 1965. http://www.presidency.ucsb.edu/ws/index.php?pid=27285 (accessed July 10, 2013).

38. MacDonald, G., et al. (1979). "The Long-Term Impact of Atmospheric Carbon Dioxide on Climate." Jason Technical Report JSR-78-07. Arlington, VA: SRI International.

39. Charney, J., et al. (1979). "Carbon Dioxide and Climate: A Scientific Assessment." Report of an Ad Hoc Study Group on Carbon Dioxide and Climate, Woods Hole, Massachusetts, July 23-27, to the Climate Research Board, National Research Council. Washington, DC: National Academies Press.

40. Weart, S.R. (2008). *The Discovery of Global Warming*. Cambridge, MA: Harvard University Press.

41. Oreskes, N. and Conway, E. (2010). *Merchants of Doubt*. New York: Bloomsbury Press.

42. Ibid.

43. Ibid.

44. Goldenberg, S. (2015). "Exxon Knew about Global Warming More Than 30 Years Ago." *Mother Jones*, July 9. http://www.motherjones.com/environment /2015/07/exxon-climate-change-email (accessed February 8, 2017).

45. Beder, S. (2011). "Corporate Discourse on Climate Change," pp. 113–129 in *The Propaganda Society: Promotional Culture and Politics in Global Context*, Gerald Sussman, ed. New York: Peter Lang.

46. Goldenberg, "Exxon Knew about Global Warming."

47. Fischer, D. (2013). "Dark Money and the Climate Change Denial Effort." *Scientific American*, December 23. https://www.scientificamerican.com/article/dark-money-funds-climate-change-denial-effort/ (accessed February 8, 2017).

48. Johnson, B. and Israel, J. (2012). "Exposed: The 19 Public Corporations Funding the Climate Denier Think Tank Heartland Institute." *Think Progress.* http://www.bing.com/search?q=funding+climate+denial&src=ie9tr (accessed February 8, 2017).

49. Leiserowitz, A., et al. (2015). "Climate Change in the American Mind: October 2015." Yale Program on Climate Change Communication. http://climatecommunication.yale.edu/publications/more-americans-perceive-harm-from-global-warming-survey-finds/ (accessed November 22, 2017).

50. Hmielowski, J.D., et al. (2013). "An Attack on Science? Media Use, Trust in Scientists, and Perceptions of Global Warming." Public Understanding of Science. *Sage Journals*, April 3.

51. Coleman, P.T. (2011). "Climate Change, Partisanship and Conflict: What's a Weather Beaten Nation to Do?" *Psychology Today*, October 30.

52. Intergovernmental Panel on Climate Change. http://www.ipcc.ch/ (accessed July 17, 2013).

53. Ibid.

54. The Kyoto Protocol. http://unfccc.int/kyoto_protocol/items/2830.php (accessed July 17, 2013).

55. Rahm, D. (2010). *Climate Change Policy in the United States.* Jefferson, NC: McFarland and Company.

56. Frontline Climate of Doubt Timeline. http://www.pbs.org/wgbh/pages/frontline/environment/climate-of-doubt/timeline-the-politics-of-climate-change/ (accessed July 23, 2013).

57. Ibid.

58. Congressional Budget Office. (2008). "Policy Options for Reducing CO_2 Emissions." http://www.cbo.gov/sites/default/files/cbofiles/ftpdocs/89xx/doc8934/02-12-carbon.pdf (accessed July 15, 2014).

59. Ibid.

60. Executive Office of the President. (2013). "The President's Climate Action Plan." http://www.whitehouse.gov/sites/default/files/image/president27sclimateactionplan.pdf (accessed July 17, 2014).

61. Ibid.

62. Rogers, K. (2015). "McConnell Tells Other Nations Obama Can't Fulfill Emissions Reduction Pledge." *Energy Guardian.* http://energyguardian.net/mcconnell-tells-other-nations-obama-cant-fulfill-emissions-reduction-pledge (accessed April 21, 2015).

63. Davenport, C. (2015). "Nations Approve Landmark Climate Accord in Paris." *New York Times*, December 12. https://www.nytimes.com/2015/12/13/world/europe/climate-change-accord-paris.html?_r=0 (accessed February 15, 2017).

64. Mann, M.E. (2016). "I'm a Scientist Who Has Gotten Death Threats. I Fear What May Happen under Trump." *The Washington Post*, December 16. https://www.washingtonpost.com/opinions/this-is-what-the-coming-attack-on

-climate-science-could-look-like/2016/12/16/e015cc24-bd8c-11e6-94ac-3d32
4840106c_story.html (accessed February 17, 2017).

   65. Ibid.

Chapter Four

   1. Bateman, C. (2010). "A Colossal Fracking Mess: The Dirty Truth Behind the New Natural Gas." *Vanity Fair.* http://www.vanityfair.com/news/2010/06 /fracking-in-pennsylvania-201006 (accessed February 20, 2017).

   2. Ibid.

   3. ibid.

   4. Ibid.

   5. Ibid.

   6. Ball, J. (2011). "Exxon Says 'Fracking' Safe as Industry Mounts Defense." *The Wall Street Journal,* May 26. https://www.wsj.com/articles/SB1000142405270 2304520804576345522519486578 (accessed February 20, 2017).

   7. Chintamaneni, V. and Showalter, J.M. (2016). "Fracking Debate Moves into Insurance Realm." *The National Law Review,* September 20. http://www .natlawreview.com/article/fracking-debate-moves-insurance-realm (accessed February 22, 2017).

   8. Botkin, D.B. (2010). *Powering the Future.* Upper Saddle River, NJ: Pearson Education.

   9. Ibid.

   10. Ibid.

   11. Katusa, M. (2011). "The Fracking Controversy." *Safe Haven.* http://www .safehaven.com/article/21049/the-fracking-controversy (accessed February 27, 2017).

   12. Maugeri, L. (2010). *Beyond the Age of Oil.* Santa Barbara, CA: Praeger.

   13. Ibid.

   14. Katusa, "The Fracking Controversy."

   15. Maugeri, *Beyond the Age of Oil.*

   16. Ibid.

   17. Ibid.; Earth Works Action. (2011). "Hydraulic Fracturing 101." http:// www.earthworksaction.org/issues/detail/hydraulic_fracturing_101 (accessed December 5, 2014).

   18. Zoback, M., Kitasei, S., and Copithorne, B. (2010). "Addressing the Environmental Risks from Shale Gas Development." Worldwatch Institute Briefing Paper 1 (Washington, DC: Worldwatch Inst.). http://blogs.worldwatch.org/revolt /wp-content/uploads/2010/07/Environmental-Risks-Paper-July-2010-FOR -PRINT.pdf (accessed February 21, 2017).

   19. Lindheim, J. (1989). "Restoring the Image of the Chemical Industry." *Chemistry and Industry* 15, 491–494.

   20. Beder, S. (2002). *Global Spin.* Cambridge, UK: Green Books.

   21. Ibid.

22. Katusa, "The Fracking Controversy."

23. Bruzelius, N. (2010). "EPA Turnaround: Collecting Data on Fracking Risks Might Be a Good Idea." https://www.ewg.org/kid-safe-chemicals-act -blog/2010/03/epa-turnaround-collecting-data-on-fracking-risks-just-might-be -a-good-idea/ (accessed February 28, 2017).

24. Ibid.

25. U.S. Environmental Protection Agency. (2015). "EPA's Study of Hydraulic Fracturing for Oil and Gas and Its Potential Impact on Drinking Water Resources." https://www.epa.gov/hfstudy (accessed February 28, 2017).

26. Warrick, J. (2015). "Major EPA Fracking Study Cites Pollution Risk but Sees No 'Systemic' Damage So Far." *Washington Post*, June 4. https://www.wash ingtonpost.com/news/energy-environment/wp/2015/06/04/fracking/ (accessed February 28, 2017).

27. Katusa, "The Fracking Controversy."

28. Bruzelius, "EPA Turnaround."

29. Skrapits, E. (2011). "Environmental Watchdog Outlines Fracking Risk." *The Citizens' Voice*, March 8. http://www.citizensvoice.com/news/drilling/environ mental-watchdog-outlines-fracking-risks-1.1115694#azz1JR6rMJa1 (accessed February 28, 2017).

30. Osborn, S.G., Vengosh, A., Warner, R.W., and Jackson, R.B. (2011). "Methane Contamination of Drinking Water Accompanying Gas-Well Drilling and Hydraulic Fracturing." *Proceedings of the National Academy of Sciences* 108:20, 8172–8176.

31. Ibid.

32. Ibid.

33. Howarth, R.W., Santoro, R., and Ingraffea, A. (2011). "Methane and the Green-House-Gas Footprint of Natural Gas from Shale Formations." *Climate Change* 106, 679–690.

34. Ibid.

35. Shindell, D.T., Faluvegi, G., Koch, D.M., Schmidt, G.A., Unger, N., and Bauer, S.E. (2009). "Improved Attribution of Climate Forcing to Emissions." *Science* 326, 716–718; Shires, T.M., Loughran, C.J., Jones, S., and Hopkins, E. (2009). *Compendium of Greenhouse Gas Emissions Methodologies for the Oil and Natural Gas Industry.* Washington, DC: URS Corporation for the American Petroleum Institute; Jamarillo, P., Grifin, W.M., and Mathews, H.S. (2007). "Comparative Life-Cycle Air Emissions of Coal, Domestic Natural Gas, LNG, and SNG for Electricity Genera- tion." *Environmental Science and Technology* 41, 6290–6296.

36. Fischetti, M. (2013). "Groundwater Contamination May End the Gas- Fracking Boom." *Scientific American*, September 1. https://www.scientificamerican .com/article/groundwater-contamination-may-end-the-gas-fracking-boom/ (accessed March 1, 2017).

37. Ibid.

38. Llewellyna, G.T., Dorman, F., Westland, J.L., Grievec, P., Sowers, T., Humston, E., and Brantley, S.L. (2017). "Evaluating a Groundwater Supply

Contamination Incident Attributed to Marcellus Shale Gas Development." *Proceedings of the National Academy of Sciences of the United States of America* 112:20, 6325–6330.

39. Chow, L. (2017). "Fracking Caused 6,648 Spills in Four States Alone, Duke Study Finds." EcoWatch, February 21. http://www.ecowatch.com/fracking -spills-duke-study-2276074733.html?utm_source=CR-FB&utm_medium =Social&utm_campaign=ClimateReality (accessed March 1, 2017).

40. Ibid.

41. Patterson, L.A., et al. (2017). "Unconventional Oil and Gas Spills: Risks, Mitigation Priorities, and State Reporting Requirements." *Environmental Science and Technology*, February 21. http://pubs.acs.org/doi/full/10.1021/acs.est.6b05749 (accessed March 1, 2017).

42. Ibid.

43. Shahan, Z. (2011). "Oklahoma Earthquakes and Fracking." Planetsave, November 7. http://planetsave.com/2011/11/07/oklahoma-earthquake-fracking/ (accessed July 24, 2012).

44. U.S. Geological Survey. "Induced Earthquakes: Myths and Misconceptions." https://earthquake.usgs.gov/research/induced/myths.php (accessed March 1, 2017).

45. Ibid.

46. Speer, M. (2012). "Fracking Risks Are Too High for Nationwide Insurance." iSustainableEarth, July 14. http://www.isustainableearth.com/energyefficiency /fracking-risks-are-too-great-for-nationwide-insurance (accessed July 14, 2012).

47. Ibid.

48. Jackson, R.B., et al. (2014). "The Environmental Costs and Benefits of Fracking." *Annual Review of Environment and Resources* 39, 327–362.

49. Ibid.

50. Public Health England. (2014). "Shale Gas Extraction: Review of the Potential Public Health Impacts of Exposures to Chemical and Radioactive Pollutants." gov.uk, June 25. https://www.gov.uk/government/publications/shale -gas-extraction-review-of-the-potential-public-health-impacts-of-exposures-to -chemical-and-radioactive-pollutants (accessed March 3, 2017).

51. Reynolds, L.R. (2016). "Seven Million Americans Could Experience Man-Made Earthquakes from Fracking." *Natural News*, April 7. http://www.natural news.com/053576_fracking_earthquakes_USGS.html (accessed March 3, 2017).

52. Ibid.

53. Loki, R. (2015). "8 Dangerous Side Effects of Fracking That the Industry Doesn't Want You to Hear About." *Alternet*, April 28. http://www.alternet.org /environment/8-dangerous-side-effects-fracking-industry-doesnt-want-you -hear-about (accessed March 7, 2017).

54. Natural Resources Defense Council. (2014). "Report: Five Major Health Threats from Fracking-Related Air Pollution." December 16. https://www.nrdc .org/media/2014/141216 (accessed March 7, 2017).

55. Loki, "8 Dangerous Side Effects of Fracking."

Chapter Five

1. Linder, D.O. (2000). "Speech on the Occasion of the 75th Anniversary of the Scopes Trial." July 10. http://law2.umkc.edu/faculty/projects/ftrials/scopes /evolut.htm (accessed March 14, 2017).

2. Ibid.

3. History.com. (2017). "This Day in History: Monkey Trial Begins." http:// www.history.com/this-day-in-history/monkey-trial-begins (accessed March 15, 2017).

4. Linder, "Speech on the Occasion of the 75th Anniversary."

5. History.com, "This Day in History."

6. Conkin, P.K. (2001). *When All the Gods Trembled: Darwinism, Scopes, and American Intellectuals.* Lanham, MD: Rowman & Littlefield.

7. Ibid.

8. Newport, F. (2014). "In U.S., 42% Believe Creationist View of Human Origin." Gallup. http://www.gallup.com/poll/170822/believe-creationist-view-human -origins.aspx (accessed March 16, 2017).

9. Branch, G. (2017). "Anti-Evolution." National Center For Science Education. https://ncse.com/news/anti-evolution (accessed March 16, 2017).

10. Ibid.

11. Ibid.

12. Ibid.

13. *Epperson v. Arkansas* 393 U.S. 97 (1968).

14. *Edwards v. Aguillard* U.S. 578 (1987).

15. Young, C.Y and Largent, M.A. (2007). *Evolution and Creationism.* Westport, CT: Greenwood Press.

16. Ibid.

17. Behe, M. (1996). *Darwin's Black Box.* New York: Free Press.

18. *Kitzmiller, et al. v. Dover Area School District* Case No. 04cv2688.

19. Linder, D. (2004). "The Vatican's View of Evolution: The Story of Two Popes." http://law2.umkc.edu/faculty/projects/ftrials/conlaw/vaticanview.html (accessed March 22, 2017).

20. Ibid.

21. Ibid.

22. Ibid.

23. Catholic Herald. (2014). "Pope Francis's Comments on the Big Bang Are Not Revolutionary. Catholic Teaching Has Long Professed the Likelihood of Human Evolution." *Catholic Herald,* October 31. http://www.catholicherald .co.uk/commentandblogs/2014/10/31/pope-franciss-comments-on-the-big-bang -are-not-revolutionary-catholic-teaching-has-long-professed-the-likelihood-of -human-evolution/ (accessed March 22, 2017).

24. Masci, D. (2014). "5 Facts about Evolution and Religion." Pew Research Center. http://www.pewresearch.org/fact-tank/2014/10/30/5-facts-about-evolution -and-religion/ (accessed March 22, 2017).

25. Ibid.

26. Shtulman, A. and Valcarcel, J. (2012). "Scientific Knowledge Suppresses but Does Not Supplant Earlier Intuitions." *Cognition* 124, 209–215.

27. Ibid.

28. Shtulman, A. (2013). "Epistemic Similarities Between Students' Scientific and Supernatural Beliefs." *Journal of Educational Psychology* 105, 199–212.

29. Ibid.

30. Shtulman, A. and Schukz, L. (2008). "The Relation Between Essentialist Beliefs and Evolutionary Reasoning." *Cognitive Science*, September. http://onlineli brary.wiley.com/doi/10.1080/03640210801897864/full (accessed March 29, 2017).

31. Barr, J. (1985). "Why the World Was Created in 4004 BC: Archbishop Ussher and Biblical Chronology." *Bulletin of the John Rylands University Library of Manchester* 67, 575–608.

32. Dalrymple, B. (1991). *The Age of the Earth.* Stanford, CT: Stanford University Press.

33. Mooney, C. and Kirshenbaum, S. (2004). *Unscientific America.* New York: Basic Books.

34. Fraley, R. (2016). "Why Science Denialism Is Costing Us a Fortune." *GMO Answers Forbes Blog*, February 22. http://www.biotech-now.org/food-and-agriculture /2016/02/why-science-denialism-is-costing-us-a-fortune (accessed April 7, 2017).

35. Ibid.

36. Brulle, R.J., Carmichael, J., and Jenkins, J.C. (2012). "Shifting Public Opinion on Climate Change; An Empirical Assessment of Factors Influencing Concern over Climate Change in the U.S., 2002-2010." *Climatic Change*, February.

37. Coleman, P.T. (2011). "Climate Change, Partisanship and Conflict: What's a Weather Beaten Nation to Do?" *Psychology Today*, October 30.

38. Jelen, T.G. and Lockett, L.A. (2014). "Religion, Partisanship, and Attitudes Toward Science Policy." Sage Open. http://journals.sagepub.com/doi/abs/10.1177 /2158244013518932.

39. Freeman, P. K. and Houston, D. J. (2011). "Rejection of Darwin and Support for Science Funding." *Social Science Quarterly* 92, 1150–1168.

40. Stewart, K. (2012). "How the Religious Right Is Fueling Climate Change Denial." *The Guardian*, November 5. http://www.alternet.org/environment/how -religious-right-fueling-climate-change-denial (accessed April 7, 2017).

Chapter Six

1. Gregor, M. (2006). *Bird Flu: A Virus of Our Own Hatching.* New York: Lancaster Books.

2. Nicolson, J. (2009). "The War Was Over—but Spanish Flu Would Kill Millions More." *The Telegraph*, November 11. http://www.telegraph.co.uk/news /health/6542203/The-war-was-over-but-Spanish-Flu-would-kill-millions-more .html (accessed April 21, 2017).

3. Gregor, *Bird Flu.*

4. Crosby, A.W. (2004). *America's Forgotten Pandemic: The Influenza of 1918.* Cambridge, UK: Cambridge University Press.

5. JAMA, Editorial (1918). "The Epidemic of Influenza." 71:13, 1063.

6. Ibid.

7. Zelicoff, A.P. and Bellomo, M. (2005). *Are We Ready for the Next Plague?* New York: American Management Association.

8. Ibid.

9. Ibid.

10. Ibid.

11. Hutchinson, R.C. (2016). "Preparing for a New Pandemic with an Old Plan." *Domestic Preparedness*, December 7. https://www.domesticpreparedness.com/health care/preparing-for-a-new-pandemic-with-an-old-plan/ (accessed April 26, 2017).

12. Department of Homeland Security. (2014). "DHS Has Not Effectively Managed Pandemic Personal Protective Equipment and Antiviral Medical Countermeasures." https://www.oig.dhs.gov/assets/Mgmt/2014/OIG_14-129_Aug14.pdf (accessed April 26, 2017).

13. Schneider, R.O. (2009). "H5N1 Planning Concerns for Local Governments." *Journal of Emergency Management* 7:1, 65–70.

14. Mossad, S.B. (2007). "Influenza Update 2007-08: Vaccine Advances, Pandemic Preparation." *Cleveland Clinic Journal of Medicine* 74:12, 884–894.

15. World Health Organization. "Influenza." http://www.who.int/csr/disease /avian_influenza/en/ (accessed May 7, 2008).

16. Schneider, "H5N1 Planning Concerns for Local Governments."

17. Rivera, R. (2006). "Prepare for Pandemic, Localities are Warned." *Washington Post.* February 25, 2006 http://www.washingtonpost.com/wp-dyn/content /article/2006/02/24/AR2006022401802.html (accessed November 28, 2017).

18. The Conference Board (press release). www.conferenceboard.org/utilities /pressPrinterFriendly.cfm?press_ID=2917 (accessed June 18, 2007).

19. World Health Organization, "Influenza."

20. Neergaard, "State Pandemic Preparations Vary Widely."

21. The Conference Board (press release); Kokjohn, T.A. and Cooper, K.E. (2007). "The Shadows of Pandemic," in *Global Epidemics* (Mari, C. Ed.). New York: H.W. Wilson, 87–97.

22. Wurtz, R.M. and Popovich, M.L. (2002). "A Framework for Supporting Disease Detection in Public Health." White Paper: Animal Disease Surveillance, Scientific Technologies Corporation, March 2002.

23. Disease Control Priorities Project. (2008). "Public Health Surveillance: The Best Weapon to Avert Epidemics." www.dcp2.org (accessed September 11, 2008).

24. Shute, M. (2007). "Spreading Its Wings," in *Global Epidemics* (Mari, C. Ed.). New York: H.W. Wilson; Zelicoff and Bellomo, *Are We Ready for the Next Plague?*; Gregor, *Bird Flu.*

25. Gregor, *Bird Flu.*

26. Ibid.

27. Ibid.

28. American Public Health Association. (2007). "APHA Opinion Survey on Public Health Preparedness." www.nphw.org/2007/pg_tools_poll.htm (accessed September 23, 2007).

29. Ibid.

30. Fischoff, B. (2005). "Scientifically Sound Pandemic Risk Communication." Briefing before the House Science Committee, December 14.

31. Ibid.

32. Thomas, J., Dasgupta, N., and Martinot, A. (2007). "Ethics in a Pandemic: A Survey of the State Pandemic Influenza Plans." *American Journal of Public Health* 97, 526–531.

33. Ibid.

34. Greenwood, J. (2015). "How Prepared Are We for Avian Flu?" *The Hill.* http://thehill.com/blogs/congress-blog/healthcare/249506-how-prepared-are -we-for-avian-flu (accessed May 10, 2017).

35. Ibid.

36. Ibid.

37. Walsh, B. (2017). "The World Is Not Ready for the Next Pandemic." *Time,* May 15.

38. Koblentz, G.D. and Morra, N.M. (2017). "Pandemics, Personnel, and Politics: How the Trump Administration Is Leaving Us Vulnerable to the Next Outbreak." *Biosecurity, Policy & Initiatives, Preparedness,* April 6. https://globalbio defense.com/2017/04/06/pandemics-personnel-politics-trump-administration -leaving-us-vulnerable-next-outbreak/ (accessed May 10, 2017).

39. Ibid.

40. Ibid.

41. Greenwood, "How Prepared Are We for Avian Flu?"

42. Walsh, "The World Is Not Ready for the Next Pandemic."

43. Gillian, K.S., Blendon, R.J., Bekheit, J.D., and Lubell, K. (2010). "The Public's Response to the 2009 H1N1 Influenza Virus." *New England Journal of Medicine.* http://www.nejm.org/doi/full/10.1056/NEJMp1005102 (accessed May 11, 2017).

44. Ibid.

45. Ibid.

46. McKenna, M. (2010). "H1N1 Lessons Learned: Vaccine Production Foiled, Confirmed Experts Predictions." University of Minnesota Center for Infectious Disease Research and Policy. http://www.cidrap.umn.edu/news-per spective/2010/04/h1n1-lessons-learned-vaccine-production-foiled-confirmed -experts (accessed May 11, 2017).

47. Steenhuysen, J. (2009). "U.S. Health Department Response to H1N1 Mixed: Study." *Reuters,* July 7. http://www.reuters.com/article/us-flu-usa-idUS TRE56669020090707 (accessed May 4, 2017).

48. U.S. Government Accountability Office (2011). "Lessons from the H1N1 Pandemic Should Be Incorporated into Future Planning." GAO-11-632. http:// www.gao.gov/products/GAO-11-632 (accessed May 11, 2017).

49. Ibid.

50. Steenhuysen, "U.S. Health Department Response to H1N1 Mixed."

51. Walsh, "The World Is Not Ready for the Next Pandemic."

52. Centers for Disease Control and Prevention. (2017). "About Ebola Virus Disease." CDC. https://www.cdc.gov/vhf/ebola/about.html (accessed May 15, 2017).

53. Putre, L. (2015). "US Ebola Response Fed Fear, Presidential Commission Says." *Medscape*. http://www.medscape.com/viewarticle/841404 (accessed May 15, 2017).

54. Walsh, "The World Is Not Ready for the Next Pandemic."

55. Ibid.

56. Greer, S.L. (2016). "Political Fights behind Uneven U.S. Zika Response." *Scientific American*, September 6. https://www.scientificamerican.com/article /political-fights-behind-uneven-u-s-zika-response/ (accessed May 15, 2017).

57. Ibid.

58. Fox, M. (2017). "Worse Than Ebola: U.S. Not Preparing for the Next Bio-Threat." *NBC News*. http://www.nbcnews.com/storyline/zika-virus-outbreak/worse -ebola-u-s-not-preparing-next-bio-threat-n753526 (accessed May 16, 2017).

59. Medscape. (2016). "The World Is Not Prepared for Pandemics." *Medscape*, November 15. http://www.medscape.com/viewarticle/871802 (accessed May 18, 2017).

60. Shinkman, P.D. (2017). "If a Pandemic Hits, the U.S. Isn't Ready." *U.S. News and World Report*, May 4. https://www.usnews.com/news/health-care-news /articles/2017-05-04/us-falling-short-on-pandemic-prevention-study-says (accessed May 18, 2017).

61. Ibid.

Chapter Seven

1. Roberts, D. (2017). "The Fake but Accurate Climate Change News Delivered to Trump? It's Fake All the Way Down." *Vox*, May 16. https://www.vox.com /energy-and-environment/2017/5/16/15645466/trump-fake-climate-news (accessed June 2, 2017).

2. Ibid.

3. Kuntzman, G. (2017). "Trump's War on Science Continues with EPA Firings." *New York Daily News*, May 8. http://www.nydailynews.com/news /national/trump-war-science-continues-epa-firings-article-1.3147062 (accessed May 22, 2017).

4. Ibid.

5. Ibid.

6. Parker, L. and Welch, C. (2017). "3 Things You Need to Know about the Science Rebellion Against Trump." *National Geographic*, January 27. http://news .nationalgeographic.com/2017/01/scientists-march-on-washington-national -parks-twitter-war-climate-science-donald-trump/ (accessed May 22, 2017).

7. Ibid.

8. Hofstadter, R. (1963). *Anti-intellectualism in American Life*. New York: Vintage Books.

9. Ibid.

10. Ibid.

11. Ibid.

12. Ibid.

13. Bauerlein, M. (2008). *The Dumbest Generation: How the Digital Age Stupefies Young Americans and Jeopardizes Our Future (Or, Don't Trust Anyone Under 30)*. New York: Jeremy P. Tarcher/Penguin.

14. Ibid.

15. Doherty, C. (2014). "7 Things to Know about Polarization in America." Pew Research Center. http://www.pewresearch.org/fact-tank/2014/06/12/7-things -to-know-about-polarization-in-america/ (accessed May 23, 2017).

16. Ibid.

17. Hetherington, M.J. and Weller, J.D. (2009). *Authoritarianism and Polarization in American Politics*. New York: Cambridge University Press.

18. Nuccitelli, D. (2017). "Trump Has Launched a Blitzkrieg in the Wars on Science and Earth's Climate." *The Guardian*, March 28. https://www.theguardian .com/environment/climate-consensus-97-per-cent/2017/mar/28/trump-has -launched-a-blitzkrieg-in-the-wars-on-science-and-earths-climate (accessed March 31, 2017).

Index

Page numbers with "t" indicate tables.

About the Author

Robert O. Schneider holds a PhD in political science and is professor of public administration at the University of North Carolina at Pembroke. His recent research and publishing efforts have been extensive in the field of emergency management (practice and policy) with a particular focus on sustainability and hazard resilience. He has also researched and written on leading policy issues where science and politics intersect. In addition to numerous peer-reviewed journal articles, he has written and published two books: *Emergency Management and Sustainability: Defining a Profession* and *Managing the Climate Crisis: Assessing Our Risks, Options, and Prospects*.